# DIY MOBILE SOLAR POWER 2020

## THE COMPLETE GUIDE TO 12 VOLT MOBILE SOLAR POWER FOR RV'S, BOATS, VANS, AND CARS

JACOB HOWELL

# CONTENTS

❀ Created with Vellum

# INTRODUCTION

Solar power systems have become increasingly popular among campers and boondockers in recent years, the latter of which rely on solar energy to run their electric cars and RVs.

This 2020 guide for a 12V solar power system will educate you on the various solar power components you need to run a solar-powered system. Through reading this book, you will learn how to plan and design a solar power system for your RV, car, and van. The planning stage will guide you through determining your solar power requirements and how you can use that information to know how many solar panels you would need to install. You will also learn how to calculate your battery size, the type of wiring needed, and wire sizing. Once you know the number of panels to install, the next step would be to come up with a layout for laying the solar panels on your roof and starting your installation process.

You will learn step-by-step process for installing various solar components and how you can activate the system to start generating electricity. You will also learn how to maintain your solar system, along with some mistakes you must avoid while handling solar-powered systems.

# MOBILE SOLAR POWER

❦

*A*s it's been for solar power systems, green energy in general has also been gaining popularity over recent years. In effect, many home and car owners have been going for an eco-friendly solar system as a mode of electricity. The solar system consists of solar panels connected to solar cells to convert sunlight into electricity. Switching to using solar energy as a mode of electricity will save you money and help you maintain the environment.

Solar power is a renewable energy source harnessed from the sun using technologies such as photovoltaics (PV), indirectly through concentrated solar power (CPS), or sometimes a combination of the two (hybrid system). In a concentrated solar power, lenses or mirrors and a solar tracking system focus a large portion of sunlight into a beam, so it can produce electricity, all while photovoltaic or solar cells use photovoltaic effects to convert light into electricity.

The electrons found in the material are stimulated by solar arrays and travel through the electric circuit, thus powering an electric device or sending the power down a grid.

Solar photovoltaics is rapidly becoming an inexpensive source of electricity for small to medium-sized applications. Photovoltaic (PV) systems rely on solar panels mounted on a rooftop or surface to convert solar energy into electricity.

## Overview of Mobile Solar Power

Mobile solar power is portable, with the solar panels mounted on an object, where it would provide solar energy. It is good to note that a mobile solar power kit can be installed in any vehicle.

Mobile solar power is used to fuel cars and other automobile devices. Solar power would convert a conventional car using electricity to run on solar energy, and it is much easier to convert an electric vehicle for use with solar power than those powered by standard gas.

Solar power will enable you to save on energy costs, as it allows you to run the car essentially for free. You would only have to ensure proper maintenance of the solar panels.

Mobile solar panels are installed to provide

onboard electricity to various components. DIY solar-powered systems will work with electric cars because the cars already have electric motors and a battery. The solar system would then charge the car battery using sunlight.

With solar power, you don't have to worry about charging your car because the solar panels installed on top can harness the sunlight wherever you go. You can maximize the amount of sunlight that reaches the solar panels by making sure the panels are kept as clean as possible. Though maximum efficiency would be achieved through the shining sun, note that you can still produce electricity during cloudy or winter months.

### Conversion of Solar Power

Solar power produces electricity by harnessing heat from the sun's rays. Photovoltaic (PV) cells, or the solar cells, convert solar energy into electricity, and the PV produces both light and electricity. Therefore, as the sun rays strike the PV material that has PV effects, it would trigger electrons inside the silicon cells, causing them to start flowing, thus generating electricity. A concentrated solar power produces electricity by utilizing solar collector devices to heat the liquid required for turning turbines connected to a generator.

This form of solar energy is converted to electricity and can power various devices or be used to power lights. The electricity can also be stored in a battery for later use. Solar cells generate a direct current (DC) form of electricity; however, the DC is converted into an alternating current (AC) using an inverter.

Based on how solar power is harnessed and used, it can either be an active or passive form of energy. Active solar energy uses special heating equipment to convert the solar energy into heat energy, that being mechanical equipment like solar cells or a solar thermal collector to generate energy.

Passive solar also uses mechanical systems to convert solar energy into heat, with passive solar heating using the energy that is collected. For example, there can be conversing energy collected by solar panels installed in a car during a hot summer day.

### STORAGE OF SOLAR Power

PV systems use rechargeable batteries to store excess electricity for later use. Most solar batteries have their own inverter integrated into the battery for energy conversion, and a higher battery capacity can store more solar energy.

Installing a solar battery together with the solar panel would ensure the excess electricity is stored for later and not sent back to the grid. If the solar panels generate more electricity than what you need, the excess electricity would then charge the battery. When there is no electricity generated by the panels —which would especially happen at night—you can use the stored electricity in the battery.

Electric vehicles that have adopted solar energy have low maintenance costs, costs of fuel, and environmental footprint. In the coming years, electric vehicles will likely take the automobile industry by storm, and the growing use of electric vehicles has increased the need for electricity to run them. As a result, there has been an increase in the installation of solar power on vehicles.

Solar power systems provide the cheap and reliable source of electricity required to run them, using the sun as a direct energy source; therefore, solar power is considered one of the cleanest, inexpensive, and reliable forms of fuel for the future.

### How Solar Panels Power a Car

Have you ever thought about how wonderful it would be to enjoy a free ride? You can have a solar-powered car that you would have to waste a dime on for fuel. Just like having a solar-powered home, cars can use the energy generated from the sun and convert it into electricity. The electricity generated will be able to fuel the car's battery and run the car's motor.

Therefore, instead of using the battery, the solar car would direct power into the electric motor. This is achieved via the use of PV cells. The PV components inside the solar panel convert the energy from sun into electricity, with the components using semiconductors made of silicon material to absorb the light from the cells. The sun rays would then trigger the electrons in the semiconductor material, creating movement of the electrons. As the electrons move, they generate electricity that powers the car's battery or motor when the car is in use.

Although most car dealer shops don't sell solar cars, many people have been building their own solar-powered cars from scratch.

## WHY USE SOLAR **Power**

- It helps protect the environment.
- It helps you eliminate the high electricity bills.
- It's more reliable.
- It helps you avoid utility inflation.

## CHAPTER SUMMARY

A 12V solar power system is a clean source of energy used to run, not only home appliances, but also in RVs. It uses the energy generated from the sun using photovoltaic cells.

A solar power system consists of multiple PV panels, which generate DC from the sun and convert it to AC power using a converter (inverter) and a racking system that holds the PV panels on your roof.

To achieve maximum solar efficiency, you can tilt the panels at an angle when installing on the roof to maximize the amount of sunlight that hits the panels.

Solar PV power is converted from DC to AC for use by other appliances. The solar energy that is generated can then power your car, van, or RV. If you love camping and boondocking, solar power systems will be ideal for your activities, as you would not only be generating electricity for direct use, but also for excess power that would be stored in solar batteries for future use when the sun is not as prevalent. The stored energy can power the LED lights in your RV, run your refrigerator, charge your laptop, and work with other appliances.

As you can see, solar power energy has a lot of benefits to individuals who love to go for weekend getaways, holidays, or camping with their RV or van.

In the next chapter, you will learn how to build your 12V solar system.

# BUILDING 12V SOLAR-POWERED SYSTEM

❧

*M*any people install solar panels as the main source of electricity or solar trickle charger to charge batteries. If a car is left idle for a long period of time, the battery will begin to lose power; therefore, it's not advisable to leave the battery for too long, as such can cause irreversible damage.

A solar trickle charger would act as a stand-alone device that generates some power to charge an RV house battery.

## 12V SYSTEM

Solar panels and off-grid systems use 12V systems as the standard voltage for off-grid electrical equipment. A 12V system is commonly used because of its inexpensive and easy to design, maintain, and equip nature.

The 12V is applicable for car electrical systems, RVs, boats, trucks, and vans, and any electrical equipment that plugs into a car or RV light socket can be powered by using ig. The equipment in this equation can include phones, laptop chargers, the music system, fridges, and cookers.

The 12V is one of the safest voltages you can use; because of its low current flow, it won't cause any electric shock.

## 12V CONCEPT:

### *Power Consumption*

Power is the basic unit of electricity and is measured in watts (W). Each 12V appliance has a label on the number of watts required to operate it; for example, laptops would need 60 watts to operate. The unit of measure helps you determine the amount of power you need to run that particular appliance.

The unit of measure can also be represented in form of time. In the above example, if your laptop needed 60 watts in 3 hours, your laptop would need a power consumption of:

$$60W * 3\ hours = 180Wh$$

### Voltage

Voltage is the ability to do something and is measured in volts. Assume that electricity is like the water flowing in order to turn a wheel—in this example, the height of the water flowing into the wheel would be the voltage. Therefore, if the water voltage is low, the water would still flow but it couldn't turn the wheels.

If you increase the voltage of the water, then it would turn the wheel and the latter can start to rotate. This concept is similar to electricity flow, as a higher voltage can make appliances operational. However, too much voltage may result in damage.

### Current

Current is the flow of electricity from point A to point B, measured in amps. Just like with the water, the amount of water flowing would represent the current. The more it flows, the more power produced. The power generated is directly proportional to the voltage and current produced:

$$Power = Voltage * Current$$

In other words, if either the voltage or current increases, the power would also increase.

. . .

## RESISTANCE

In any electricity generating system, resistance reduces the flow of electricity within a particular component or a wire. All 12V components, including connection cables, have an internal resistance. These components can consume power without doing anything useful.

## CONNECTION OF SOLAR Components

When connecting the components together, you can use a parallel or series connection. In some cases, you can use both.

If the electrical circuits are connected in series, the components will have a shared voltage and constant current. If your solar panels and batteries are connected in series, the voltage is the sum of voltage across the panel.

In a parallel connection, the current will be shared among the appliances while the voltage remains the same.

## USING Solar Panels as the Source of Electricity

Solar panels installed in your RV house system act as the primary source of power. To get enough power to run and maintain your RV system, you will have to install several solar panels and other gears. A complete solar kit has all the components needed to install and regulate power for your routine operations.

An RV solar panel requires sunlight to generate electricity, and the batteries attached to it will require a chemical reaction for them to generate electricity.

*STEPS:*

## 1 ) Power requirements

Before installing any solar cell in your RV, you have to determine its power requirements. Generally, you need to have a solar panel that generates somewhere between 200 to 400 watts in order to recharge a RV's battery.

200 watts is the *minimum* requirement for providing basic solar power needs. For example, you would need this amount of watts if you hope to produce the correct amount of electricity needed for LED lights, powering the radio, and for any propane refrigerator electronics. If you need more lighting in the RV, a television, and other electronics that require electricity, you will have to go for a 400W solar panel.

To know how much power you need for your RV

or car, you have to match the power (amps) output of your solar cells with the power requirements in your car or RV.

## 2) Understand the system design of the RV

YOU HAVE to know the basic design of your vehicle to know the system components it needs. The solar panel is always installed on the vehicle's rooftop so it can collect enough sunlight for powering the system.

## 3) Selecting the system components

SOLAR CELLS ARE CONNECTED with wires that carry the power into the charge controller. The wires are then connected to the battery. You can also use fuses to protect the vehicle's electrical system.

When choosing the components for your system, you can go for higher efficiency components along with those that can keep the cost down.

Some components for a 200W system can include:

- Two polycrystalline solar panels of 100 watts each.
- A PWM charge controller of 20 amps.

- Two fuses of 20 amps each (one fuse should be inline MC4).
- Wires with ring and MC4 connectors

Requirements for 400 watts solar system would include:

- Two or four 100W monocrystalline solar panels.
- A MPPT charge controller of 30 amps.
- MC4 to SAE adapter and a Zamp outlet for an RV.
- Two fuses of 20s amp each (one fuse should be inline MC4).
- Wires with ring and MC4 connectors.

## How Solar Energy Works

Before designing your solar system, you will need to understand how the solar cells generate electricity.

The solar panel/module is made of a layer of silicon cells aligned with a metal frame, glass casing, and a wiring system that transfers the electrical current into the silicon. The solar cells have a semi-conductor material, such as silicon, that converts the light energy into electricity—that is, the silicon has some conductivity properties that can easily absorb and convert the sunlight into reliable electricity.

As the sun's rays hit the silicon cell, it forces electrons in the silicon material to move, resulting in the flow of the current. This process is known as the "photovoltaic effect." In short, when you expose semiconductor material to the sunlight, it will create an electric current.

The photovoltaic effect is the main functionality of generating electricity through the solar panels technology.

***Steps involved:***

1. The photovoltaic silicon cell absorbs the sun's rays (electromagnetic radiation). The sun energy then moves in the form of electromagnetic radiation, such as light or radio waves, and the type of the radiation is based on the wavelength of radiation emitted.
2. As the sun's rays get absorbed by the silicon cell, they would trigger the flow of electrons in the cell. As a result, it generates an electric current.
3. The generated current is in DC, which is then converted to AC for use. In DC, the current flows in the same direction, whereas with AC, the current flows in direct directions/reverse of the direction.

## TIPS TO GETTING the Best from Your Solar Panel System

- **Device efficiency:** When installing solar applications, always consider the basic efficiency in each solar device. Install devices that require less power to run—for example, you can try using LED lights, or installing a newer flat screen television instead of an older one, the latter of which would be less efficient.
- **Device runtime:** Always turn off any devices you are not using so you can minimize your overall power usage. Lights are a good example of something you can turn off if you don't need it. Make note that less power consumption will reduce the power generated by the solar system, contributing to the device runtime.
- **Solar panel output:** Learn how much power your system produces each day, as this will help you understand your power budget for a specific time frame.
- **Capacity of your battery bank:** Knowing how many amp hours remain for your battery to drain will enable you to plan for the off-grid power consumption. Always

run heavy activities when you already
have a full battery bank.

The battery charging is determined by the
number of solar panels you have. The more solar
panels, the faster your battery bank will recharge. If
you have several batteries connected to your RV,
then your RV can run off-grid for longer.

- **Specific period of the year:** Sometimes,
  based on the time of the year, you can
  experience a longer or shorter period of
  solar charging, which can affect your
  overall power charging system. Therefore,
  how you mount the solar panels on the
  surface will determine how much power
  will be absorbed, especially during days
  with less sunlight.
- **Geographical location:** If you live in an
  area that is usually very cloudy, especially
  in the extreme north, you may need to tilt
  mount your solar panels in order to obtain
  an optimal solar input. Cloudy areas will
  impact the amount of sunlight absorbed
  by your solar panels and may make
  generating electricity a little more
  difficult.

## Benefits of Using **Solar Panels**

- You will be reducing the amount of pollution to which you contribute. Therefore, investing in solar power for your recreational vehicle has plenty of benefits, not only to you, but also to the environment. Solar power is a green source of energy and doesn't produce any pollutants to the environment.
- The electricity generated through solar panels operates without any smell or noise.
- Solar panels don't have any moving parts, but you can place them in mobile objects to help you enjoy solar power on the go.
- Solar panels require minimal maintenance.
- They are cost-effective. Though the initial capital may appear high at first, with time, it will become cost-effective. You can also learn how to tune the solar panels to help you meet your electricity needs more efficiently, helping you save time and money in the long run.
- Solar power is free; therefore, you don't have to spend on that costly fuel.
- Solar power is a clean source of energy, so in using it, you won't have to deal with any

messy oil or fuel. There are no exhaust fumes nor noise with solar panels, compared to when using a generator.

## Basic Solar Components

### Solar panels

The solar panels, or solar modules, have photovoltaic cells made of silicon material and are used to convert sunlight into electricity. The term *photovoltaic* means "electricity produced from light." These photovoltaic cells contain both positively and negatively charged silicon film under a thin layer of glass.

The protons from the sun heat the cells, knocking electrons from the silicon. The negatively charged electrons are stored in one area of the silicon cell, resulting in the creation of electric voltage. Connecting the individual solar panels in series forms a solar photovoltaic array that enables you to collect the current needed to power the vehicle. Based on the size of installed solar panels, you can have multiple arrays of photovoltaic cables that terminate in a single electrical box (the fused array combiner).

The combiner box consists of fuses that protect

each solar module and connection cables to transfer power to the inverter. The produced electricity is in DC, which is then converted to AC suitable for use by the vehicles or at home.

## INVERTER

An inverter is connected close to the solar panel. The inverters do make some noise; therefore, you should factor that in when looking for a suitable location to install them. These inverts convert the DC current generated by the solar panels into 120V of AC.

If the inverter is connected directly to a dedicated circuit breaker in the solar panel, it will allow you to use electricity immediately after conversion to AC.

There are two types of inverters: string inverters and micro-inverters. A string inverter is installed in a shaded area or wall and converts energy from several solar panels into AC electricity. A micro-inverter is installed on each solar panel on the back, where it works on each one independently.

In a string inverter, if one panel is not collecting energy from the sun, it will affect the performance of the other panels. In a micro-inverter, shading in one panel won't affect the performance of the other panels.

. . .

### ELECTRIC METER and net meter

Electric meters measure the flow of energy and direction. Once you install a solar power system, you should upgrade the electric meter to meet net metering capabilities.

In a solar power system connected to the utility grid, the inverter converts the DC power generated by the solar array into 120/240 V AC power. The generated power is stored directly into the utility distribution system; it is net metered to ensure there is reduced demand on the use of electricity from the utility when the electricity is being generated. This will help you have a reduced utility bill.

The current solar power systems have grid-tied systems that automatically turn off when the utility power goes off. The generated power from the solar system is first consumed by the current electrical loads in operation, and the remaining power is directed onto the electrical grid. When more electricity is produced than what is needed for consumption, the utility meter will turn backwards.

### SOLAR CHARGER CONTROLLER

A solar charger controller is used to regulate the amount of voltage or current flowing from the solar panel into the battery. This helps prevent overcharging of the battery while also prolonging the battery's life.

The power conditioning unit synchronizes the DC power supply converted into AC power with the electricity utility. It also protects the batteries against electric faults caused by either a short circuit or even line-to-ground fault. This can be achieved through the use of a thermal-magnetic circuit breaker.

## BATTERY BANK

A battery bank is used to store energy for use on demand. The excess energy not required at the moment to run the electrical load on operation is stored in batteries. Batteries increase the price of the solar system.

In this guide, we will focus on 12V batteries. If you have high power demand, you can use multiple 12V batteries that fit in your 12V solar system and connect them in parallel. This will ensure voltage remains 12V while the battery capacity (current) increases. Each battery is always labelled with its voltage and capacity measured in amp-per-hours (Ah). The Ah determines the battery capacity over a specified time. For example, a 40Ah battery means it runs 40A within 1 hour, or 1A for 40 hours. The higher the current drawn from the battery, the more the voltage drops due to the battery's internal resistance.

. . .

### Solar racking system

A solar racking/photovoltaic mounting system is a product that fixes solar panels securely on various surfaces like on the rooftop of a house or a car, or ground. The solar racking beneath the solar panel plays an important role in the installation process. The aluminum material used makes it lightweight; as a result, it is easy to install on rooftops compared to other materials made of metal.

Solar cells should be installed perpendicular to the sun rays. In order to maximize on the total energy production per annum, the racking system should fix solar arrays in the same angle as the latitude of the array location.

### Components of Solar Racking

There are different types of solar racking systems. Each solar racking system has different components, though some components are common to all solar racking systems. Those common components would include the following:

### Flashings

When installing solar cells on the roof, you will have to drill holes for fixing. If these holes are not covered well, they will start to leak, causing damage to your property. Flashing material is used to cover the drilled holes and prevent water from leaking, and the flashing material is fitted beneath the shingles.

There are different designs made for each roof—for example, metal roofs or roofs with tiles. Flashings are designed with a unique shape for each type of roof material.

## Mounts

You can use mounts to install solar panels on a roof securely. The mounts are fixed using a bolt passed through the flashing material into the rafter. Different racking systems can use different types of mounts; therefore, evaluating the underside of your roof ensures the rafters are structurally installed and spaced appropriately to accommodate the solar panel mount.

## Rails

The mounts are part of the solar racking system that holds up the rails, so the solar panel can sit on top of it directly. Rails are made of aluminum tracks and can be installed vertically or horizontally on the roof.

The rails allow you to run the solar system

wiring behind the rails, which helps reduce clutter and improve the safety of your DIY solar installation.

## CLAMPS

You can use clamps to secure the solar panel in place. The clamps would link the solar panel to the rails below. In any installation, you can use both mid-clamps and end-clamps; mid-clamps would be installed between the solar panels and hold them in place on both sides, whereas end-clamps are installed at the end of the solar arrays and are usually longer.

## FUSES

Fuses are very important in any electrical system installation. They protect the wires against any short circuit or power surges that may cause damage to your solar appliances or even drain your battery faster.

Ensure you have the right size of fuse that can handle the current flow, as the installed fuse should be able to handle the current draw from every connected appliance.

## WIRES

Selection of wires is very important in your solar installation process. The wire thickness selected should minimize line losses that occur as a result of internal resistance from the wire. The flow of the current would determine the line losses; that is, the higher the current, the thicker the wire and the higher the line loss.

The wires use American Wire Gauge (AWG) to determine its thickness. If the wire you are using is not thick enough, you will lose much power while it transfers from the panels to the battery. AWG thick wires are very expensive; therefore, if you're transferring via a long distance, you may consider looking into alternative wire options.

The power drop is based on the current transferred, and a higher current will result in higher line loss or power drop. To reduce line loss, you can reduce the current and increase the voltage using step ups. Installing a step up on the power source will convert the power before any transmission into high voltage and low current. At the battery end, you have to use step downs to convert the higher voltage back to its 12V normal voltage. You can then connect back your 12V appliances.

## CHAPTER SUMMARY

The 12V concept is essential when building a 12V system for your RV, van, and car. You have to evaluate the power consumption, voltage, and resistance experience when electricity is flowing from one solar component to another.

When buying a solar power system you have to consider two main important factors: is the system worth it, and *why* is it worth it?

The cost of a solar power system can be quite high, making it available for only a few people. Nevertheless, solar power will save your vehicle from experiencing much wear, maintenance, and generator fuel consumption, among other things. Based on your power needs, you can choose any type of solar panel.

You learned through this chapter how a solar power system would work, along with the various components required to power that solar power system. You also learned about solar racking and mounting equipment, which you can use to secure your solar panels on the roof. Lastly, you learned how to increase the efficiency of your solar panels.

In the next chapter, you will learn how to plan and design your solar power system.

# PLANNING AND DESIGN OF SOLAR POWER SYSTEM FOR RVS, TRUCKS, VANS, AND CAR

*A*re you a camp addict? Despite your boondocking experience, you may not have had a great portal solar experience. Installing solar panels on your RV's roof can be a great idea and will provide you with the best off-grid power solution while camping.

However, before buying the solar power system, you will need to size it. Sizing involves evaluating all electrical devices needed to run a solar power system successfully. This can be achieved by taking a closer look at your daily power requirements, then using this information to account for solar power efficiency, depth of battery discharge, and determining the actual vs. ideal solar isolation. After determining those requirements, you can then figure out how many watts of PV your solar power panels will need for battery capacity and on your roof.

The battery bank and inverter-based system will provide the power needed to run your RV system throughout, and you won't have to have a generator for your RV. The battery can power any 240V appliance, like the air conditioning, refrigeration, cooking, and washing machine.

**Design Criteria**

Your RV solar system design should depend on:

- Your average daily load demand.
- Maximum power surge demands.
- Voltage of your solar panels and power distribution.
- Total cost of installation and your budget.
- Required electrical appliances.
- Maintenance services required.
- Type of control system installed.

When designing the solar power system for not only your RV, but also your car, van, and/or boat, you will need to consider the following:

- Your solar supply strategy and how to manage the supply of generated electricity to meet the electricity needs of your vehicle.

- The distribution of electricity from the DC to AC loads.
- Solar power system voltage.
- The size of the solar panels or solar arrays.
- The size of the battery bank.
- Ratings of the major components required for solar power installation.

The solar power system voltage for small DC systems is distributed using extra low voltage (ELV) of 12V, 24V etc. The actual voltage for the solar power system will depend on the size or capacity of your solar panels and the connection distance between the battery and load points in your vehicle. Electrical appliances connected in your vehicle and the solar controller will also determine voltage rate.

### Requirements for a Solar System
The solar panel for your RV should have:

- RV-grade solar panel.
- Mounting brackets to install the RV on your roof.
- Mounting hardware required to fix the brackets on the roof.
- Wiring system that connects the solar panel to the battery and charge controller.
- Charge controller & inverter.

- Battery bank.
- Battery bank connectors.
- A fuse or breaker.
- A silicon sealant, required to permanently install the panels.

## PREPARATIONS FOR INSTALLATION

You don't necessarily have to be an expert to install a solar panel on your RV, but you will need basic experience on how to handle DC electrical and RV wirings. Nevertheless, if you don't have the minimum experience, don't worry—DIY solar panels come with manuals that can instruct you on how to go through the installation process. This guide will also teach you a step-by-step process for installing solar panels into your RV or car.

Before proceeding to the installation process, you will need to have:

- A voltmeter.
- A hammer.
- A jigsaw.
- A ladder.
- A cordless drill.
- Drill bits of various sizes.
- A stud finder.

- A willingness to read and try to understand a set of instructions.

WHEN INSTALLING a solar module on your RV permanently, you will need to drill holes on the roof. When doing so, you have to be careful and use the right solar racking tool to avoid any water leakage into your RV.

## PLANNING for Your Off-Grid Solar System

Planning your off-grid solar power system will involve a series of trade-offs. The process includes investigating how many panels you will need and the number of batteries to invest in. You will also need to evaluate the power consumption, power input from the panels, and power storage in the batteries.

## CALCULATING Your Power Consumption

### Step 1: How much power do I need?

The first step in the planning phase is to know how much power you will need to run all your appliances. You need to ask yourself: what do I need to power? Make a list of all the appliances you wish to power and check the power consumption for each appliance at the labels or under each one's instructions. The majority of these appliances will indicate their power consumption in watts (W) or in kilowatts (KW).

***Step 2: How many hours will I need to power the appliances?***

The next step is to figure out how many hours per day you will be running each appliance. This will give you the total energy consumption over a period of time. When considering the hours to power an appliance, take note of the summer and winter months. Knowing the hours to power the appliances will enable you to calculate your daily power consumption.

### CALCULATING **Power from Your Panels**

Since you already know how much power you will need, the next step will be to determine how much power you can get from your panels and how many panels you would need. If you already have panels, you can calculate how much power they will produce and what to power with them.

Based on the size of the panel, you can calculate the power generated during sun peak hours of the day and the power generated during off-peak hours.

### CALCULATING **Required Battery Storage**

Battery capacity is determined by the amount of

power produced during the winter and summer months. During the summer period, you can produce power up to 4.5 times that which you would produce in winter. If you use a certain amount of solar energy in the summer, you will probably need to use a fifth of that energy in winter; therefore, it may be a good idea to store some of the energy you generated in summer and use it in winter. You can achieve this by having a battery bank to store the energy while also reducing your power demand during the winter period. You will need to obtain a battery that satisfies your daily power needs and store enough power for days with less energy production from the sun.

The unit of measure of battery capacity is amp hours (Ah). You can convert it to watt-hours (Wh), taking the battery unit hour and multiplying by voltage. For example, if your battery capacity is 20Ah and the voltage is 12V, then the power available will be:

$$20Ah * 12V = 240Wh$$

In this, the battery can either supply 240W within 1 hour, or 120W within 2 hours. Therefore, we can see that the more energy that is consumed, the faster the batteries will drain. Note that if you drain the battery below the recommended voltage, then you will not be able to power it. Care should be taken to ensure the batteries are not discharged beyond 50%.

**DESIGNING the Solar Power System**

*Glossary of Terms*

In this topic, I will be using the following abbreviations, so you should get familiar with them. They will give you a better understanding of what I am talking about and help you recognize the concepts instead of scratching your head every time you come across a relative abbreviation.

- **Amps:** Abbreviation of amperes, which is an electrical current unit of measurement.
- **Imp**: Maximum current produced by the solar panels. It indicates the maximum number of amps produced by a solar panel in perfect conditions. Whether you're using parallel or series connection, or a combination of both wiring, Imp determines the current generated by the solar panel array.
- **Isc:** Represents short circuit current of the solar panel or the amount of current produced if both positive and negative wires of a panel are connected. The term is used more commonly when

determining the requirements for circuit protection.

- **Vmp:** Indicates the voltage of the solar panel at maximum power. That is, the voltage needed to generate imp-rated solar power. You have to figure out wire sizing from the solar panels up to the controller, and the total Vmp will depend on whether you're using a series wire connection or parallel wiring, or a combination of both.

- **Voc:** An open circuit voltage. It is the amount of voltage generated by solar panels when the sun hits directly on the panel with no wiring. This voltage is used to calculate the solar controller size.

- **String:** A series of two or more solar panels wired together, forming a series connection. You can use multiple strings of solar panels, each wired in series. The multiple strings can be wired together to form a parallel connection at the combiner box.

- **Combiner Box:** Acts as a junction box through which you can connect the lead wires together from the panel into a single pair of leads (positive and negative wires). The wires run through the solar controller, and they greatly reduce the number of wires needed to run through your RV.

- **Home run:** Wires that are supposed to run from the solar panel into the combiner box. These wires should be UV and weather-resistant since they are installed on your RV's roof.
- **AWG:** Abbreviation for American Wire Gauge, this is a standard for wire size conductors commonly used in North America. The wire size diameter depends on the AWG number. A small AWG number indicates a large wire size diameter.

*WIRE SIZING*

Wiring your solar power system transfers the power generated down into the RV through the solar controller, then into the battery. The layout of the solar components and the length of your RV determines how long the wiring runs.

If you have a long wiring system between the solar panels and the battery, it will result in voltage loss because more time was spent transferring electricity down the long wiring, thus some power that was generated was lost due to the long wiring that electricity had to travel through.

You can reduce the amount of power lost by using large-sized (gauge) wiring.

**Note:**

- If you plan to use a large-sized (gauge) wire, be prepared to spend more money because the copper material used for making quality wiring is very expensive.
- Installing large-sized wires could also be more difficult because the thick wires will need more space, making it tougher for the wires to run through a hole.
- The diameter of a gauge wire is heavy; therefore, you will also need to consider the weight of large-sized wires when choosing a wiring system for your RV.

If you know what type of wiring you need for each portion of solar installation in your RV, you can then use a voltage drop calculator.

### *Voltage Drop Calculator*

To use a voltage drop calculator, you will have to understand:

- The voltage (Vmp) of the solar panel, which is calculated based on whether you have a series or parallel wire connection.

- The maximum solar current power (Imp), measured in amps for the solar panels connected, either in series or parallel.
- The wiring distance from the solar panels to solar controllers, then to the battery.

A voltage drop calculator would estimate the drop in voltage of an electrical circuit. This depends on the wire size used, wiring distance, and any anticipated load current.

The calculator operates with the assumption that the circuit is operating under normal conditions; that is, operating at room temperatures with normal frequency. The voltage drop is based on the current condition of the wire, frequency, temperature, and even the connector used. In any circumstance, the voltage drop should be below 5% under a fully loaded condition.

### What Causes a Voltage Drop

When an electrical current flows through the wires, it has to pass through a contrary pressure level called *impedance* in an alternating current. In a direct current, the pressure level is *resistance*.

For example, when water is flowing through a hose pipe, it requires a certain amount of pressure to push the water through that pipe. You consider voltage in electricity and the water flowing through the pipe as being the current. The hose pipe would cause a certain level of resistance, which depends purely on the thickness and shape of the pipe. Wires work in the same manner—the type and size of the wire determines the level of resistance.

If the circuit experiences an excessive voltage drop, such will result in a loss of efficiency, thus leading to a flickering of lights, poorly heated room, and sometimes burnout because the motors can become hotter than their normal temperature. Selecting the right wires for your recreational vehicle ensures the drop doesn't go beyond 3%.

A voltage drop can be caused by:

1. **Inappropriate material used in the wires**: Using wires made of copper material, rather than aluminum, will reduce voltage drop in a given length of wire size. Copper is one of the best conductors and will have less of a voltage drop.
2. **Wire size**: The size of the wire will determine the voltage drop. A large wire size gauge (large diameter) will have less of a voltage drop compared to smaller wire sizes with the same length.
3. **Wire length**: If you have short wires running through the RV, then you will experience less voltage drop compared to

when using long wires with the same size and/or diameter. Voltage drop is a problem when running long wires while installing solar panels on the roof; therefore, you should use an appropriate wire gauge when running wires in a long distance.

4. **Amount of current flowing**: The amount of current flowing through the wires also determines voltage drop levels. A high amount of current flowing through the wires increases voltage drop. The capacity of the current moving through the wires is similar to "ampacity." Ampacity is the maximum number of electrons that are pushed at once via the wire.

Ampacity in wires is affected by the insulation of the wires. If the temperature is high, the wires will become too hot. The material used in making the wires and the speed at which AC flows also both affect the ampacity.

### Layout of the RV Solar System

Before buying any solar components, you need to know the function of each component and how much power you need for all your needs. This will help you determine the number of solar panels you need and how many can fit on your RV's roof. You will also need to know how much wiring you need to fix all the components together.

## Mapping the RV Roof

*Image credit voyagerix/shutterstock*

You need to map out the roof of your trailer to know how much space you have and the shape of available space. Do to this, you can climb up the roof of your trailer with tape and a paper pad to take these measurements. You can make a general sketch of the shape of the RV roof as well, and get an exact location of any component on the roof that may hinder the installation of the solar panel. This can be a TV antenna, air conditioner, roof vent, or any other component on the rooftop. Measure the distance from the edges and also from the front and rear of the RV. From this, you can determine the total usable space you can use to install the solar panels. Once you have taken all the measurements, you can then transfer all the information on a clean sheet of paper. Graph paper will probably help a little more in mapping the roof.

You can draw the exact scale where you want the solar panels to be installed, along with any other components, on the paper. This will make your work easier when you start installing the solar panels. You can also add all other existing components on the roof and the exact scale to use based on your measurements and installing placements for them on the roof. This will provide you with a scale of how your RV roof will look, and will also let you know of the challenges you may run into while installing the solar panels and how to avoid them.

You can have a sample of your solar panel placement on the roof by cutting out the mockups scale and placing them on the exact area to install the components. You should always put into consideration how shading from taller objects, like TV antennas and air conditioners, can affect the amount of power produced by the solar panels.

## Determine How Many Solar Panels and Where to Place Them

There are different things you can do to determine the power requirements for your RV, which will enable you to know the number of solar panels you would need to install. One method you can use is to go camping for a few days without the use of an AC generator. Then, you would determine how long it will take you to run the RV from the stored charge on a normal electricity usage.

### RV SOLAR SYSTEM *Example 1*

Imagine you had two batteries and, within two days, they drained up to 50% with the batteries, providing you with energy of about 200Ah. Each battery supply powers at a rate of 100Ah. With that, you can calculate how much solar power you need to run your RV on a single day. In this case, the energy consumption when you camped for two days (100/2) is 50Ah of energy on a regular day.

When using batteries to power all your electricity needs in the RV, avoid discharging the battery to below 50%, as this may shorten the battery life. With the information on the storage capacity of your batteries, you can determine the number of solar panels needed to replace the 50AH of energy consumed on an average day.

During sun peak months in summer and spring, you can have an average of 5 peak-sun hours on a daily basis, which can make your batteries full within a short time.

A solar panel with 100W power can generate an average of 6A/peak-sun hour. Therefore, a single solar panel should be able to produce, on average, 30Ah daily. In this case, you will only need two solar panels of 100W to fully recharge your battery and run your RV.

Once you figure out how much solar power is needed to run your RV, you will need to search for the best solar panels to use and a battery bank. Always determine the specifications designed especially for your RV solar panel systems.

### RV SOLAR SYSTEM *Example 2*

If you have a 300Ah battery bank that can last for about three days, then it will be best to recharge the battery daily to avoid draining it. Calculate the amount of watts you need to replenish per day based on peak-sun hours.

Assuming you have a 95W solar panel and there are 5 peak-sun hours in a day. If your power consumption is 1/3 of 300 = 100Ah per day, then you will have to replenish 100A within 5 hours. Each hour, you will have to generate 100Ah/5h = 20 Amps.

If you're using a 95W panel kit, then you will need to have 4 solar panels with 20A of charging at a rate of 5.4A/panel: 20/5.4 to get 3.7, or simply 4 solar panels. The per-day charging rate will be based on the weather conditions, time of year, and your camping location.

Adding more solar panels to your RV can recharge its batteries faster. A charger controller installed on the battery will ensure you don't over-charge the batteries.

### FACTORS TO CONSIDER When Checking Specs for an RV Solar Panel System

#### 1. Watt rating

Before buying each solar panel, you will have to confirm the watt rating for each panel and the total watts to size the system accurately.

#### 1. Peak power (amps)

THIS IS the maximum power in amps generated from the solar panel when there is full sunlight exposure. The power is measured in amps, and some solar panels have a 5A peak power rating. With a sun-peak of 6 hours, you can enjoy up to 30Ah charging power.

If you need a charger controller installed, make sure you know the total combined peak power ratings from the solar arrays, so you can buy a charger controller with an accurate amp rating.

### 1. **Peak power (volts)**

PEAK POWER IS the maximum power that the solar panels can generate when there is full sunlight, measured in volts. Sometimes, solar panels tend to drop the output up to 2V, which can affect the efficiency and charging rate of the solar panels.

During low-light periods, it affects the charging rate. Therefore, when choosing your solar panel specs how the low-light will reduce the charging rate.

### 1. **Tolerance**

YOU SHOULD CONSIDER the solar panel tolerance since it has an impact on the amount of power supplied by the solar panels. There are cases where the solar panel would produce less power by 3% or sometimes more, by 3%. Thus, the panels should have a tolerance of -3% and +3%. Knowing the % tolerance for each solar panel will help you determine the amount of power needed every day accurately.

. . .

**Other Purchase Considerations for Portable Solar Panels**

### 1. Waterproof solar controller

Always make sure you're buying a waterproof solar charge controller; otherwise, if you buy solar panels without a waterproof controller, you will always have to take it out of wet weather.

### 1. Electrical connectors

You need to know what type of electrical connectors your solar system uses, though this can be changed later on. Sometimes, you can use wiring harnesses and alligator clips that can allow you to connect directly to the battery.

Manufacturers of portable solar panels always have recommendations for which connectors to use and how to program the solar controller. If your recreational vehicle has a pre-wired solar port that allows you to plug solar panels in, then examine what type of connectors are needed before purchasing any. Alternatively, you can change the ends of the connector.

If the panel needs to be installed further away from your rig, then you will have to invest in more extension cables. Depending on the type of connectors in your system, you can buy extensions that are compatible with your system.

Common recommended connectors include:

- MC4 connector.
- Anderson connector.
- SAE connector.
- No connector—some portable solar panels have small gauge wires with alligator clips at the end. This allows you to only extend the wiring of the panels.

### 1. **Solar panel extension cables**

PORTABLE SOLAR PANELS can be placed anywhere where there is full sun. You can also put them at an angle that points directly at the sun throughout the day.

These solar panels are ideal in areas where there are lots of trees, as they can help keep your RV away from trees that shade the panels. You can buy the extension cables to connect the solar panels, and the best extension cables with the correct size or gauge wire with multiple length options will all have a MC4 connectors at the end.

. . .

## Budget Versus Your Solar Needs

Solar panels are the greatest way for going off-grid using your RV, and being aware of your power requirements on the RV will enable you to pick the solar panels that meet your needs. The more solar panels you need, the more money you will also need.

The number of watts in the solar panels determines how fast your batteries will be recharged, along with the overall output of the system. Since more solar panels need more money, you would need to get enough solar panels that you can afford without straining the savings in your bank. Get as much solar power that will enable you to meet your minimum needs, and try to live within your means. If you need three panels and can only afford one or two, just buy what you can afford; you can always expand later when you have more money.

## Spacing of Solar Panels

Buying more solar panels for your RV means having more space to install them. More than one panel will require more wiring, a bigger controller, and a good installation plan. Installing large solar arrays would be done once the RV is parked, which will allow you to have a good orientation of the panels directly toward the sun. It also allows you to spot any shaded spots and ensure the panels are installed away from the shading area.

Once you know the solar panel requirements, you can go ahead and install it onto your rooftop. Assuming you need 600W of solar, you can install 3 panels with 200W each. The panels can be selected based on your RV's roof size. For example, you can have two rectangle panels of 200W each, and a square panel with 200W. This ensures the panels fit your RV roof design properly.

Always be prepared to be flexible when installing solar panels. Sometimes, things may pop up that you never would have anticipated.

### INSTALLING **Wires from the Roof into the RV**

You can run the wires into the battery and drill a single hole to allow them to pass. Alternatively, if there is a refrigerator vent in the RV roof, you can use it to your advantage by passing the wires through the vent.

Assuming you have two rectangle solar panels to install in your RV, you can use MC4 "Y" connectors to connect the wires from the solar panels into a single set that can run from the roof of the RV down to the batteries via the fridge vent. This can reduce the number of wires on the roof going down to the battery.

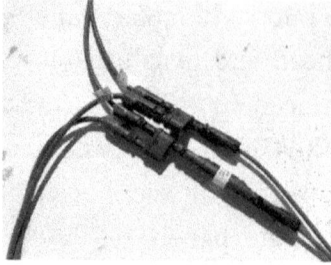

FOR EXAMPLE, you can connect four wires from the solar panel into two sets of wires using MC4 connectors.

**WHERE TO INSTALL Solar Controller**

In most cases, solar controllers are placed next to the RV batteries. You can install the solar controller in the forward baggage compartment of your RV, which is also close to the batteries. Installing the controller in the baggage compartment will protect from harsh weather conditions. You can also install it on the tongue of your RV.

**12V RV BATTERY**

A recreational vehicle will house a 12V solar power system that would supply all your electricity needs. The 12V battery is designed to store a large amount of energy required by your RV. The energy supplied is enough to allow all electrical appliances and other devices in your RV run effectively.

It will be difficult to run your appliances without the 12V RV battery. Though there are alternative sources of energy, like wind energy, 12V batteries are used widely to boost or quick start the energy requirements of your RV. Having the battery on your side, you can enjoy all the comfort and ride anywhere.

When deciding on the battery tank, make sure you get a battery that stores at least twice the amount of power you spend within a day. This will ensure your batteries last longer.

### How to Store RV Battery When Not in Use

Keeping the battery in good shape and storing it properly will extend its lifespan. If you don't plan to use your battery for a long period, you should know how to store it safely.

RV batteries self-discharge if left for long without use. Though this depends on the type of battery installed and the current weather conditions, you still need to learn how to store your battery and prevent the battery safely so it won't discharging itself and become unresponsive. If this happens, it can become unresponsive and will be unable to recharge fully.

You can store the battery in the following ways:

- Before storing the battery, recharge it to 100%, then disconnect it from the RV. This protects your battery from discharging and reduces parasitic loads such as TV antennas, alarms, stereos, and power boosters, among others.
- You should wipe the top of the battery to remove any residue matter that can contribute to discharging of the battery.
- Look at the battery's electrolyte level and replenish the distilled fluid as often as possible. The battery's water level should be maintained at 90% to avoid the battery expanding when charged.
- The battery should be stored in temperatures of above 32°F (0°C), or temperatures higher than 80°F (27°C). If the battery is frozen, don't recharge it because it will explode.
- The battery should be placed on higher ground, away from water and potential flooding.
- If the battery is to be stored for a long time, it should be checked regularly and recharged once a week.

### How to Charge **the RV While Driving**

If you don't recharge your recreational vehicle, it can ruin your day, as you will be unable to unleash the maximum potential of your solar panels and the benefits they offer.

Sometimes, solar energy is unavailable; therefore, you should learn how to recharge the RV battery while driving. This will enable you to have a better camping or road trip experience. To recharge the battery, connect the RV battery to the truck's power source, which is also a battery using high quality cables and plugs. Due to the huge amount of energy transferred from the truck's battery to the RV, high quality cables and plugs are recommended.

The cables will be more reliable and will provide a more stable connection. As a result, they reduce voltage drop while maximizing the RV battery charging process. Furthermore, when traversing a rough terrain, the high-end plugs will ensure the battery is connected securely, allowing you to not have to worry about any disconnection while driving in rough areas. If your RV battery is not connected securely to your truck's battery, your RV battery could drain, effectively ending your camping trip quickly.

You can insulate your batteries from short circuits by using rubber boot slips on the battery breakers. A 50A automatic circuit breaker can be plugged on the positive terminal to prevent any possibility of a fire hazard.

You should always monitor the RV battery to avoid overcharging it. If you suspect you have over-charged the battery, examine the level of electrolytes in the battery and, based on the results obtained, you can add distilled water if needed. During extreme weather conditions, you should monitor your batteries more often.

**Battery Monitor**

A battery monitor is a must-have tool in any RV installation. The monitor will help you know whether the batteries are fully charged within the day and how much power you use each day. That is, it monitors the power in and out of the battery.

The monitor also keeps track of the battery voltage and temperature, especially if you have a sensor attached. The monitor passes a message to the solar charge controller on the status of the battery via Bluetooth technology. When installing a battery monitor in your RV, always ensure there is no load between the battery and the shunt. A load between the shunt and battery will prevent the shunt from reading the amount of power, in or out. The load will also not be recorded by the battery moni-tor. To enable recording of the load, you would have to install a distribution block and shift the ground wires.

### Wiring of the RV

Wiring of the solar panels will be the most important step in your solar installation. The wires are used to pass the generated power from the solar panels to the solar controller, then the battery bank thereafter.

Using wires with a small gauge (diameter) or wires without copper strands limits the flow of electricity through the wires. This affects the flow of power, causing your solar panels to not function as expected.

Always choose a wire gauge that gives you less than 2% voltage drop throughout the entire solar system. You can calculate the amount of voltage drop by using a wire size calculator. To be able to obtain less than 2% of voltage drop, you need to have a large wire gauge (diameter). The wire gauge depends on the length of the wire needed to install the panels and the total power (amps) generated by the solar system. For example, if you have two rectangular solar panels, you can use the 4 AWG wires combined from two solar panels and connect them to the solar controller.

The majority of solar controllers accept a maximum of 6 AWG wire size, which has a smaller diameter than the 4 AWG. From the controller to the batteries, you can also use 4 AWG wire or 2 AWG wires.To know the exact wires your RV will need, use a voltage drop calculator. When you have a long RV or you have a large solar array (series of solar panels), it is best to determine the wiring size.

### Calculate Wire Size Based on Maximum Power Capacity

You should calculate wire size based on the maximum capacity your solar controller can handle. This is essential if you're considering expanding your solar power system in the future. If there is not enough room on the roof for solar panels, you can also consider having portable solar panels to help you boost your power production.

The portable panels will require you to run a ground deployment through the solar controllers. You won't need to have two solar controllers, as they may work against each other. When doing wire size calculations, you should also factor in the total watts from the ground deploy panels.

*Example:*

If your truck has a rooftop of 24ft. and you need 600W panels to satisfy all your power needs, the wiring size used for solar installation will be:

- 10 AWG wires running from the solar

panels to the combiner.
- 6 AWG wire running from the combiner wires to the solar controller.
- 6 AWG wires running from the solar controller into the RV batteries.

You can use a welding cable, which is a high-quality copper wire and more flexible for your 6 AWG wires, and water and corrosion-resistant solar cables for the 10 AWG wire runs.

If your RV is longer, you will end up with more solar panels and long wiring runs to install the solar panels. Therefore, using a voltage drop calculator to determine an appropriate wire sizing is ideal.

## Use Quality Wires

Use welding cables to fix the combiner box securely with the solar controller and the solar controller to the batteries. These wires have a large size gauge (diameter), which is needed for the connection.

Welding cables are highly flexible and are usually made of copper strands. To wire the solar panel with the combiner box, you should use UV weather and corrosion-resistant solar cables. Tray cables are highly recommended due to their weather-resistant nature. There reason behind needing UV and corrosion-resistant cables is because the wiring runs on the roof of your RV and will need to withstand UV rays that come from the sun.

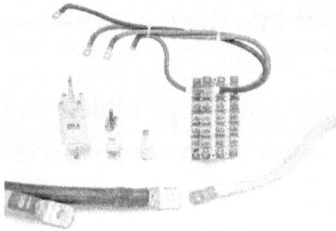

*Image source: Marcel Derweduwen/shutterstock*

DEPENDING on the amount of power generated, you can install a circuit breaker between the solar panels and controllers. If the solar panels can't produce more than 30Ah, you can use a 30A circuit breaker. You can also use a 40A circuit breaker between the solar controller and the batteries since the solar controller would have a maximum of 30A.

You will also need to install a catastrophic fuse next to the battery tank, which will protect your wires in a case where there is a short circuit on the positive terminal wiring from the battery. If no catastrophic fuse is installed, your RV's power bank can produce an amperage output within a short period. Sometimes, the battery may have some sparking or arcing, which a catastrophic fuse can prevent. Most RVs come with a catastrophic fuse and, based on its capacity and location, it may be able to sustain your RV battery needs. If the capacity of the fuse is not enough to protect the battery wires, you will need to replace it.

COMBINER BOX

Each solar panel has positive and negative wires. The wires run down from the roof of the RV to the rig and, eventually, to the solar controller. If you have three solar panels to install, instead of having six wires running all the way from the roof into the solar controllers, you can use the combiner box to combine the wiring together and form a single pair of large wire gauges. Thus, you will have few wires running on your RV.

The combiner box can consolidate the six wires from the three solar panels into two wires. Wires used to connect the solar panels into the combiner box should be small gauge wires, while those running from the combiner box into the solar controller should be large gauge wires.

Combiner boxes are always fitted on the roof at a central point where all the wires are combined together to form a single pair of wires. It also helps in reducing the length of wires running from solar panels into the combiner box and makes it easy to drill a single hole on the roof where the wires from the combiner box will pass through down into the rig, then into the solar controller.

### Types of Combiner Boxes

There are two types of combiner boxes: pre-made or your own designed combiner box. To make your own combiner box, you will need a power distribution block put in a weatherproof box. Along with that and for this type of box, ensure the holes drilled on it are also weatherproof.

If you have small rigs, you can place the combiner box inside. You don't have to obtain long wires that connect the combiner with the solar panels. Installing combiners inside will eliminate the need for a weatherproof combiner. In this case, you will only need to have a weatherproof method that ensures the wires run inside the RV. If you pass the wires through the refrigerator vent, you can place the combiner box inside. If you install it on the roof, make sure you get a weatherproof combiner box.

A recreational vehicle can house a 12V solar power system that will supply all your electricity needs. The 12V battery is designed to store the large amount of energy required by your RV. The energy supplied should be enough to allow all electrical appliances and other devices in your RV run effectively.

It will be difficult to run your appliances without the 12V RV battery. Though there are alternative sources of energy, like wind energy, a 12V battery is widely used to boost or quick start the energy requirements of your RV. Having the battery on your side can help your trip be more comfortable and will allow you to ride anywhere.

### CHAPTER SUMMARY

In this chapter, you learned how to plan requirements for installing a quality solar power system in your RV. Thus, based on your daily power demand and load size, you can now install a solar panel that meets your needs. You also need to consider factors, such as time of operation in correlation with your daily power load, installation cost, maintenance cost, and your specific preferences of the solar system when making your plan.

The installed system should be able to satisfy the design criteria and be both cost-effective and reliable. Economic evaluation of the system is very important to compare the cost versus benefits obtained. Getting familiar with the resources needed in a solar power system's installation enables you to do it yourself. If you have little knowledge on handling solar system equipment, don't worry— some DIY solar tools have manuals for installing them, making it easy for you to do it yourself.

When planning on your solar needs, always opt for the maximum solar installation. You may start with a small solar array based on your budget, but expand your system in the future. In short, factor in your future solar needs when planning and designing your solar system. For example, if you install a 400W solar panel in your RV and size all the connection cables and the solar controller to only handle 400W, you will be unable to expand your system in case you want to have 600W instead. In other words, you will be left with little room for expansion; in case you did want to expand, you would be forced to buy new cables and replace your solar controller to accommodate that 600W. This may turn out to be more expensive for you in the end.

You should only incur the expenses at once, along with running the cables in your RV once. Therefore, when planning, ensure there is room for future expansion of solar power. Also, consider having ground deploy panels as a way of expanding your solar power. The quality of the solar power installation is essential and should be worth every coin spent, especially if you love to do a lot of off-grid camping.

In the next chapter, you will learn about the RV solar charge controller, solar panel selection, and using parallel and series solar panel connections.

# RV SOLAR CHARGE CONTROLLER

*T*he solar charge controller will be the center of any solar installation process you do. It converts the energy generated by solar panels on the rooftop (and ground deploy in case you're using them) into a form that can recharge your RV batteries, while also supplying 12V power to run all the equipment in your recreational vehicle. The amount of power needed to run your RV will depend on the number of equipment installed and their power consumption rate. Installing solar power systems on your RV will provide you with a clean, renewable, and sustainable source of energy.

Just like any other system, solar energy is not 100% efficient. There are times when losses do occur; therefore, it is good practice to take precautions that will help reduce any losses that may occur. Using a solar charge controller will be great for improving the efficiency of your solar system. If the solar controller is not installed in your RV, it may result in a deficiency of energy production. A charge controller installed close to the battery will help prevent overcharging or undercharging, both of which could make the battery burn out or run dry.

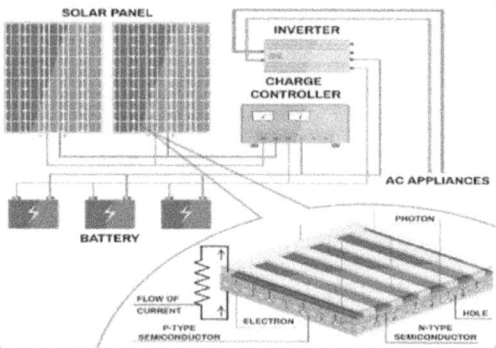

*Image source: sisVector/shutterstock*

AN RV solar charge controller controls the charging and discharging of the batteries with energy generated from the solar panels and any other connected load. A solar charge controller is an essential component in solar power systems for a recreational vehicle. It is also known as the battery regulator or charge regulator because it regulates the voltage and current flowing in and out of the battery.

. . .

## TECHNOLOGIES USED

- Maximum power point tracking (MPPT)
- Pulse width modulation (PWM)

When using an MPPT solar charge controller, the voltage and current generated through the solar panels are measured in real time, and the maximum power generated is supplied to the RV batteries, which depends on the capacity. A PWM solar charge controller uses switches and solid state gats to supply a constant current directly to the battery, though the input from the solar panels will vary.

PWM is the cheapest technology and is used more commonly with smaller power generation panels. However, MPPT is more concerned with real-time current and voltage monitoring, and the technology is more efficient compared to the PWM. Both MPPT and PWM solar charger controllers will be essential components of the solar panel system installed in your RV because of the RV' stand-alone nature. In other words, a solar charge controller will be an important component in the RV solar power installation because the RV is a mobile power consuming entity. As a result, you won't need to connect the solar panels to a grid electricity supply.

The solar charge controller makes RV energy much more viable and safer.

## How to Use a Solar Charge Controller in an RV

*Image source: styleuneed.de/shutterstock*

For any solar charge controller to be fully functional, it must have:

- An RV solar battery.
- A solar/off-grid inverter.
- Transmission cables.
- Solar panels.

A solar array or panels generate electricity when exposed to sunlight. The semiconductor material or silicon absorbs the electrons from the sunlight and transfers them inside the silicon material. The electrons transferred generate DC, which is eventually transmitted into the charge controller.

The solar charge controller is connected to the battery bank and transfers the DC to the battery at a specific voltage. The battery then passes the DC to an inverter connected to it to convert the DC into AC, connected to the load. Transmission of electricity from the panels up to the inverter involves the use of cables and wires. Thus, the solar charge control acts as the brain of the entire RV, with the electricity generated via the solar power system. It manages the discharging and recharging process in the battery while tracking the charge transmitted to the battery and regulating the voltage and current in the battery based on the RV's requirements.

## Sizing the RV Solar Charging Controller

You can charge your vehicle's battery quite easily with a 12V RV battery and a 12V solar charge controller. You can also size the solar charge controller easily for your RV.

If you have a 200W solar panel, you would need to use a 20A solar controller; likewise, if you have installed 400W panels, you would need to have a

40A solar charge controller. The solar charge controller rating is given by RV solar power from the panels divided by the voltage of the RV battery:

$$Amps = Watts/Volts$$

If you're using any ground deploy solar panels, they should also be factored when calculating the solar charge controller rate. The voltage rating should also be a factor in selecting the solar charge controller. The controller should have the capacity to hold the maximum voltage it can handle.

Depending on whether you're using parallel or series connection, such will help you determine the actual voltage rate to use in sizing the solar charge controller. As stated, beyond the voltage, the solar charge controller selected should also be able to handle maximum amp input generated from the solar panels (both rooftop and ground deploys). Solar charge controllers designed for RVs are optimized to meet the RV solar applications needed, such as size, function, volume, and heat dissipation.

### Tips for Choosing the Best RV Solar Charge Controller

### Maximum Power Point Tracking (MPPT)

Maximum power point tracking (MPPT) is one of the best solar charge controllers out there, and it helps you get the maximum output from your solar panels while increasing its battery life. It improves the battery performance by boosting the charging rate and reduces the charging time of the batteries. Having one is an added advantage, especially when you are camping and unsure of the weather conditions.

**Cooling**

Heat dissipation is very important when choosing solar controllers. Solar charge controllers are usually mounted on the RV passenger compartment. Though it seems the best placement for installation of the controller, it can also bring problems, especially if you have a controller with more than a 30A rating.

The heat generated should be dissipated; otherwise, the charge controller will definitely fail. Therefore, when buying controllers, you should look for those with good heat dissipation. If you get the ones with enhanced cooling fans, that will be even better.

GROUNDING

Every vehicle has a negative ground system, so if you choose an MTTP solar charge controller for your RV, then you should also buy the negative ground charge controller along with it.

### Selecting Charge Controller

Always choose a solar charge controller that is compatible with your battery voltage (either 12V, 24V, or 48V). Installing an MPPT solar controller will automatically detect the battery system voltage.

To choose the right solar charge controller, you have to calculate the maximum current and voltage output generated from all connected solar arrays. Based on the results you obtain, select a solar controller that exceeds those obtained ratings. Choosing a solar charge controller with higher ratings will allow for an expansion of solar arrays in the future.

You can also use a programmable MPPT solar controller that can extract maximum amount of solar power easily from the connected solar panels. The controller is compatible with different solar panels and battery bank configurations. It will als allow you to reprogram it later whenever you decide to upgrade your solar panels or batteries.

The MPPT solar charge controller ratings usually represent the maximum current output in form of amps, and not the maximum input current from the solar panels. For example, If you have a 100V/30A MPPT solar controller, it can accept voltage from solar panels of up to 100V and 30A output from a charging battery.

Always buy an MPPT controller that controls the output current automatically. When the output current is exceeded during a sunny day, the MPPT controller limits the amount it accepts. In normal charge controllers, if the charge exceeds the maximum output current, it shuts down and generates an error code.

### WHEN TO INSTALL a Solar Charge Controller

You would install a solar charge controller whenever you install solar panels in your RV. The controller is mounted inside the RV and next to the battery or in conjunction with the solar panel. It can also be installed next to the battery inside the RV. In some vehicles, the battery is mounted on the outside; therefore, you can attach the solar charge controller outside the RV with the battery. In other situations, solar controllers are installed as a different component away from the battery or solar panel, but have cables that connect the controller with the battery and the solar panels.

The solar charge controller makes solar energy that is generated more reliable and efficient. Even when the vehicle is moving, the solar power system will still be in working mode, thus producing more electricity. Since the solar controller is an important component of your RV, it should be kept away from moving parts or equipment that generates heat.

The installation of the solar system controller ensures there is minimal transmission losses in RV stand-alone systems, especially in cases where the electricity generated is limited for consumption by the connected load. Therefore, installation of solar charge controllers will depend on the architecture and interior design of the RV.

## Solar Controller Sensors

The MPPT solar controller is considered a smart controller in the market because it can adjust the charging mode based on the voltage from the battery; that is, the length of time it will take to charge depending on the battery condition in terms of its temperature and the voltage on a specific day. The solar controller relies on sensors to know how much it should charge the batteries. This helps the controller determine the status of the battery, which would be its voltage and temperature.

If the solar controller doesn't have these sensors, you can install a Victron solar controller followed by a Victron battery monitor. The Victron battery monitor will act as the sensor for determining the battery voltage and the temperature.

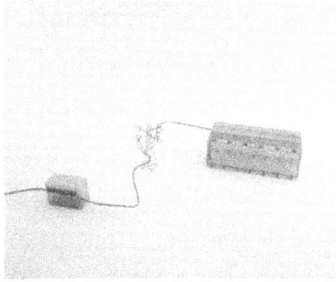

*Image source: Italy3d/shutterstock*

THE MONITOR WILL ALSO establish communication with the solar controller via Bluetooth connection to ensure the controller knows what is happening to the battery, as well as record the battery's temperature and voltage. Thus, the solar controller should be in a position to compensate for any voltage drop between the battery and controller. It will know whether the battery is still charging or when the battery temperatures are high.

### How the Battery Type Affects Solar Charging

Nowadays, lithium batteries are commonly referred to more often for use in RVs and vans, compared to traditional lead acid batteries. Lithium batteries have low internal resistance while charging, whereas lead acid batteries have a high internal resistance. A high internal resistance can limit the amount of current accepted by the battery while in charging mode. Lead acid batteries slow down the solar output because most of the time, when the lead battery is in its charging phase at 80%, the solar energy generated at that period is often wasted.

The current accepted by the battery tapers off for some hours until the lead battery is 100% full. As a result, it reduces the solar energy produced since all the power demands of the battery bank have to be met. Thus, during the sun peak hours of the day when solar power is in abundance, you can power your electronic devices like laptops, Instapot, etc and consume less battery charge.

Lithium batteries charge at a faster rate than the traditional lead acid batteries. Due to the low internal resistance experienced with the lithium batteries, more energy is absorbed from the solar panels, making the battery 100% in a very short time. They use a steady charging rate, just like when pumping a gas, and will continue at the same rate until full; therefore, the switch to using lithium batteries will guarantee you faster charging.

## SOLAR PANEL SELECTION

Solar panels come in different shapes, sizes, design styles, and power output. When selecting solar panels, many people usually want all of the panels in their solar power system to match, including ground deploys; however, doing so should depend on whether you plan to have a series or parallel connection, or even a combination of both.

If you're using different solar arrays with different watts, you should:

- Match the solar panel watts rating if you have series connection of two or more solar panels. This increases the overall solar voltage. If the cables are wired in series, each solar panel will have its own amperage (Imp rating) and it is reduced to have the lowest Imp rating of the string.
- If the connection is parallel, then you can mix and match the solar panel watts rating. In this case, the solar panels should have the same Vmp rating.

As a general rule of sizing with a solar array, you should have 1W of solar power per amp hour of battery size. For example, if you have 100W of solar, you should have an equivalent of 100Ah battery bank. You can add one more solar panel to the system to account for any losses in efficiency of tasks.

The roof space of your RV is the limiting factor in the installation of solar panels. Camper vans have less space compared to RVs, so you have to determine the maximum number of solar panels that fit on the rooftop of your vehicle. In case you don't have enough space on the rooftop, you may take into consideration the use of portable solar panels. Always think about the sizing of the battery bank. Portable solar arrays can help you to recharge the battery fully.

## SOLAR EFFICIENCY

A solar panel system is used to convert energy from the sun into electricity. The total electricity generated through the sun arrays is referred to as the solar panel efficiency rating, which ranges from 15% to 25%. Therefore, a single solar panel with high-end solar cells would produce 25% electricity from the sun's energy under normal conditions. The sun releases the remaining energy as heat energy.

## FACTORS THAT AFFECT the Efficiency of Solar Panels

- The average peak hours of sun each day.
- The sun's angle. This is usually affected by your geographical location, time of the day, and year.
- Solar shading, either from tall trees, roof-

mounted antennas or air conditioners, and buildings.
- The type and quality of the solar cells installed. Solar panels with monocrystalline cells are more efficient compared to other cells, while thin film panels are great for absorbing indirect sunlight.
- Weather conditions.
- Temperatures in the atmosphere. When it's very hot, less energy is produced.

Monocrystalline solar panels are ideal for RVs, cars, and vans because they produce the highest amount of power per square foot. Given the limited roof space in vans and RVs, these panels are highly recommended. As mentioned, monocrystalline cells have higher efficiency compared to the other types of solar panels (polycrystalline and thin film).

Polycrystalline are also available if you're on a tight budget and you have a lot of space in your RV. These panels are slightly larger compared to mono panels and are less tolerant to heat. They are cheaper compared to mono panels, however, and can also perform well under certain conditions.

Thin film is the cheapest among the three, with a low sunlight efficiency and high heat tolerance. They take up a lot of space, hence why they may not be ideal for RV installations.

. . .

## Why Solar Energy Efficiency is Important for RVs

When moving from one area to another with your RV, it can be difficult to maintain a maximum efficiency level from your solar panels. Such can be achieved when they are installed in buildings or houses because it's easier to predict the sun's location throughout the year and tilt the solar panels at an angle that will absorb maximum sunlight throughout the day. In a RV, however, this is not possible because you're rarely stationary unless you decide to camp only in one specific location.

Most of the time, your location and environment will continue to change as you move from one camping site to another, thus making it difficult to achieve maximum efficiency with mobile solar panels installed on your RV.

### Solar Panel Voltage

At some point, you may want to install solar panels of a higher voltage of 28+V (Vmp rating). The higher voltage panels are very large and require extra room to be accommodated on your vehicle rooftop.

An alternative would be to have lower voltage panels that are wired in series to give you a higher voltage. However, this requires you to have solar panels with the same amount of watts. If you have an MPPT installed solar controller, it will perform well with a voltage input between 2 and 2.5 higher than the normal 12V battery.

### SERIES AND PARALLEL **Solar Connection**

Sometimes, you may wonder whether you should use a series or parallel connection. Each configuration has its own advantages and disadvantages. The series or parallel configurations affect the voltage and current (amps) from the solar panels.

When connecting more than one solar panel, you will have to decide whether to connect the cables in series or parallel, or sometimes a combination of the both. When using series connection, you have to add the voltages from the panels together, while the current will remain the same. If you're using a parallel connection, you will have to add the currents together while the voltage remains the same.

It will be easy to combine the amperage (Imp) rating in a **parallel** connection, but it will be impossible to combine the voltage (Vmp). The voltage depends on the lowest Vmp rating of each individual solar cell; therefore, it's always important to match the Vmp rating to your panels closely.

If the solar panels cables are connected in **series**,

the voltage (Vmp) of both panels will combine, but the panels' Imp will not combine. The amperage depends on the lowest Imp rating of each panel. For example, if you connect 160W panels with a 240W panel in series, the amperage will reduce to 160W. The remaining 80W in the 240W panel will be a complete waste.

Connecting the panels in series increases the voltage rating in a solar array, but there is no difference on the amperage. As a result, you will be able to use a small wire gauge for the home run, or even the same wire gauge size to make the home run cables longer. If you're using MPPT solar controllers, then you need to have higher voltage inputs.

## Series Connection

In series connection, every panel is connected to the next, forming a string of panels. The negative terminal from one panel is connected to the positive terminal on the next, and so forth. Connecting solar panels in series will allow you to generate a higher voltage using low current. You can sum up the voltage from each panel to obtain the overall voltage in the solar array.

### *Pros*

1. You can use inexpensive thinner cables to connect the solar panels to the solar controller. This saves you money and effort because thick cables are expensive and difficult to work with.
2. It is easy to make and understand. Series connections are of a simple design that is easy to understand and connect.
3. It is easy to expand. You can add more panels to the connection and increase the voltage output.

*Cons*

1. It is prone to high voltage drop. If one panel fails or gets shaded, it will affect the entire solar array.
2. As you increase the number of panels in the connection, the greater the resistance in the connection.

PARALLEL CONNECTION

In a parallel connection, each panel is connected to a centralized wire from the RV roof. It uses a combiner box, which uses a single wire to connect all the wires from the positive terminal of the panels and another wire to connect all the wires coming from the negative terminal.

In a parallel connection, the voltage of all the panels remains the same, but the amperage is the sum of the overall amperage for each solar panel. If you connect your solar panels in parallel, you will be able to generate a high amount of current at a very low voltage. As more panels are added to the system, the more current will be generated.

*Pros:*

1. It is affected less by shading compared to other connections; if one panel is shaded or fails, it doesn't affect the entire solar power system. It only reduces the power output generated.
2. There is equal voltage for each panel connected in the array.
3. You can connect or disconnect a panel without affecting the functioning of the entire system.

*Cons:*

1. It requires the use of thicker cables, which can be quite expensive.
2. You need a combiner box to enable you to transfer generated electricity from the solar panels to the solar controllers.
3. If you pass the same current throughout the circuit, the connection will fail.
4. You can't increase the voltage in a parallel connection.

### Traditional vs. Semi-Flexible Solar Panels

In the past, the traditional glass solar panel was the only available solar panel in the market. Introduction of flexible and semi-flexible solar panels in the last few years has taken the market by storm. With them, users no longer have to rely on traditional glass panels, but can also choose to use either flexible or semi-flexible solar panels on their sailboats, RV applications, or cars.

Each solar panel style has different characteristics that meet specific needs. Both panels perform well based on the application they are designed for, so it's on you to choose the style based on your planned uses.

### Traditional Solar Panels

Traditional glass solar panels have different styles and sizes. They have been in existence for more than 70 years, and compared to other types of solar panels, they offer the best protection of the solar cells from bad weather and any impact.

Each solar cell and connection wire is encapsulated into an epoxy resin beneath the glass material and the solar cells have different configurations for different solar panel sizes. For example, an individual solar panel uses a DC battery system rated as either 12V or a 24V. You can combine several panels to achieve a specific solar output.

### CHARACTERISTICS *of traditional solar panels*

1. **Higher efficiency**: The panels use monocrystalline cells that provide maximum solar efficiency.
2. **Impact resistance**: The impact resistance in this type of panel is very high. As a result, it provides more protection to the solar cells.
3. **Heavy**: A single 100W glass panel weighs 16 lbs alone. If you decide to use this type, make sure your RV roof can handle this weight plus the mounting hardware.
4. **Generates heat at the bottom**: The cooling of traditional glass panels takes

place underneath the panels, thus releasing more heat at the bottom than what is released on the glass top. Therefore, when installing solar panels, leave some gap under the panel to allow heat to escape.

5. **Cable and combiner box location**: Cables from the solar panels and the combine box are always mounted on the back of the panels.

6. **Rack-mount or mounting bracket**: RV solar panels require at least L-shaped brackets to mount the metal frame securely on the RV rooftop. You can also use a custom rail system to mount the panels on the roof. These components add weight to your solar installation and may contribute to future flexibility. Depending on the option you go for, the installed panels should withstand heavy winds, especially while driving above 60 mph.

7. **Inaccessible sections of the roof**: Due to the size and styles of these panels, they can make some sections of the RV roof inaccessible.

SEMI-FLEXIBLE SOLAR PANELS

A semi-flexible solar panel is lightweight with a thin design compared to traditional glass panels. A single 100W semi-flexible panel weighs about 3 to 5 lbs. Just like with traditional panels, semi-flexible panels also have different shapes and design configurations.

The output of each panel is compatible with a DC battery system of either 12V or 24V. You can also combine the panels to obtain a certain output level. They're more fragile and more prone to damage if not handled well during shipping.

The panels were initially designed to be installed in sailboats, but they can be mounted on slightly curved surfaces. Over the years, they have become more popular for RV solar power installation because of their flexibility, low weight, and portable nature. These panels perform just like any other traditional glass panel.

### CHARACTERISTICS *of semi-flexible panels*

1. **High efficiency**: They use monocrystalline cells to provide you with maximum solar efficiency.
2. **Impact resistance**: Semi-flexible panels have a thin layer of material that covers the solar cells; therefore, you can damage them easily if you drop or walk on them.

The cells have a high-quality back-contact that increases the durability of the cells.

3. **Light-weight**: A single 100W panel can weigh a maximum of 5 lbs.

4. **Flexible up to 30 degrees**: Flexible panels can be mounted on curved surfaces because it can curve up to 30 degrees. Bending more than 30 degrees may result in micro cracks, which can reduce the efficiency of the cells.

5. **Cable and combine box location**: The combine box and cables should be installed on the top of the panel, in case you're having a flat mount.

6. **Heat release**: Flexible panels consist of a thin layer at the top, which enables them to release the generated heat both on both the top and bottom layers. When mounting the panel, leave a small gap on the bottom to allow cooling, especially during hot weather.

7. **Maintain a low profile**: These panels are mounted directly on the RV roof; therefore, they're invisible from the ground level.

8. **Provides room to access the roof**: Compared to traditional glass panels, flexible panels are easier to move,

especially when mounted on the flat surface of the RV roof.

9. **Offer multiple mounting options**: These panels have grommets fixed at the ends to make it easy for you to fasten them on a solid surface. You can also use double-sided tape to secure the panels on the roof permanently or use a rack similar to the one used in securing traditional glass panels. Always make sure the mounted panel can withstand 60 mph of wind or more while you're driving.

### SHADING of Solar Panels

Whenever you select the location for installing solar panels on your RV roof, you need to consider shading. A roof air conditioner and TV antenna can shade your solar panels, limiting the power output generated. Trees can also shade the solar panels, so don't pack your RV under a tree and expect it to generate a lot of power.

The connection of the solar panels, whether in series or parallel, will determine how the shading of the panels will affect the entire power production system of your RV. If the panels are connected in series and one of the panels is shaded, it will affect all other panels connected. If you have a parallel connection and one of the panels is shaded, it will only reduce the power output. Therefore, having a series connection and MPPT controller increases the total voltage from the solar system, but any slight shading will affect your entire system. Parallel connection has few problems with shading effects, as it only reduces the overall power output without increasing the total solar panel voltage.

If you have an even number of panels, like 2 or 4, you can use a combination of series and parallel connections to maximize on solar voltage while reducing the shading effects.

When connecting the solar system in series, you have to match the amperage (Imp) rating of the panel. That is, using panels with the same wattage in a series of solar arrays/strings. The strings' Imp rating will be the lowest Imp for a single solar panel in the string.

If you use a combination of series and parallel connections of panel cables in a solar array, then you have to apply the parallel connection rules at the combiner box. This is because each string of panels will be connected in parallel at the combiner.

## PARALLEL CONNECTION RULES

- The total Imp current results from adding each panel in a string.

- The voltage (Vmp) of a combined solar panel string limits the lowest Vmp rating of the entire string.
- Always make sure the Vmp of each string is the same. If one string has a lower Vmp, it will affect (bring down) the voltage of all the other strings, thus reducing the Vmp to the lowest Vmp rating of the strings.

## WHAT SIZE and Shape to Install

Solar panels are made of different shapes and sizes. For example, most RVs and cars use 100W solar panels with a 12V battery bank. You can combine the panels in different configurations to obtain a specific voltage or get a certain current output. Most panels are rectangular in shape, although some manufacturers may provide custom-made shapes for specific applications.

When determining the size of your solar panels, you need to evaluate your RV roof. Ask yourself—how many can fit on the roof and still give me a space to move around on my rooftop? You also have to determine whether there is any other equipment on the roof that may cast some shade on the solar panels. Lastly, you need to figure out where to run the cables from the roof into the RV.

The size of your solar charge controller will also

determine the amount of solar power that needs to be generated. If you buy an MPPT solar charge controller that accepts more voltage, then it's a guarantee to have more solar configurations. If your MPPT charge controller accepts higher voltages, then you can even install solar panels that accept 24V battery systems, such as 170W and 200W solar panels.

**Depending on the panels you choose to use and your power needs, it's highly recommended that you stick with the panels that have the same voltage output and a particular current specification.** This will help you eliminate losses or any other complications that may occur due to a connection of different solar panels.

### Portable Solar Panels

If you prefer portable solar panels as an alternative, you won't have to worry about learning much of the technical stuff. Portable 100W and 200W panels come with connection cables and a charge controller to enable you to charge your RV battery from the sun directly.

You would only need to unfold the panels and place them in a position facing the sun. Connect the cables from the panels to the solar charge controller and the RV battery to charge. When done with charging or you're ready to move, you would only need to fold them up and store them safely in your RV.

If you love boondocking, portable solar panels can allow you to add an additional on-demand solar power to your existing solar panel system in your RV. Portable solar panels are designed for easy setup and use.

## Chapter Summary

A solar charge controller is an important component of mobile solar power systems in an RV or any off-grid solar power connected device.

Using a solar system connected to batteries will give you a good return on investment, as the solar controller keeps the batteries safe and more efficient. Not having a solar controller in your solar system can be catastrophic to the life and operations of your RV solar batteries. Any back up battery in the RV should be connected to the solar charge controller to maximize solar panel output and extend the lifespan of your solar power system.

In the next chapter, you will learn how to install and wire solar panels, mount solar controllers, and connect to the battery.

# INSTALLATION OF RV AND TRUCKS
## SOLAR SYSTEM

❧

To build your own solar panel customizable to your needs, you would only need to have basic solar power system know-how. Installation on an RV/truck requires an appropriate laid down plan to enable you install the solar panels effectively.

Your installation plan depends on the type of solar panel you have and its size (either flexible or traditional glass panels), your roof size and shape (is your RV or trailer roof flat, curved, or made of fiberglass or rubber?). You also need to consider where the battery bank and solar controller are located.

## Roof Preparation

After determining your power needs based on all appliances installed in your RV, you can determine the total wattage you need to install to satisfy your daily consumption. Mapping your rooftop will also help determine the number of solar panels to fit in your RV; for example, if your power needs dictate that you should make 600W from the solar panels, you can decide to buy three solar panels of 200W each.

When buying solar panels, keep in mind the size of the RV rooftop. Depending on the roof size, you can decide to use either rectangle or square-shaped solar panels. The initial mapping of the rooftop will enable you to select the appropriate shape for your solar panels.

In this tutorial, we will focus on installing two rectangle 100W panels that accept a 12V battery system. If there is any other object on the rooftop, like a TV antenna, that may affect the installation process or occupy a lot of space, remove it first or install it in an area where it won't interfere with the installation of the solar panels. A hole left after removing an antenna or any other object on the roof should be covered or filled with a sheet of aluminum.

### FINALIZE the Layout of the Rooftop

You should come up with the layout for how you plan to install the panels on your RV's roof. The layout should show the exact location for where each solar panel is to be installed based on its shape. The panels should be located as far as possible from any other object installed on the roof to avoid shading. You should also leave some space where you can walk in between the panels in case you need to get on the roof later.

### Solar Installation Safety Tips

- As a general rule, you should always review all safety precautions included in the solar kit manual before starting any installation.
- When the panel is exposed to sunlight, don't touch any of the active electrical components, such as the terminals.
- Avoid installing the panels under inclement weather. If the weather conditions are not favorable, do not install solar panels, especially if you are installing them outdoors.
- Do not step or sit on the solar panels during installation.
- During installation, cover the panels with either a box or cloth to prevent charging during the installation process.

## Installation Process

Once you figure out the exact location for each panel, you can mount the solar panels on the roof of your RV or truck. The mounting process depends on the type of panels you're using. For example, rigid or traditional panels will require you to use screws to fix the panels securely on the roof. Some panels may require you to use adhesive to mount them on the roof. Therefore, based on the roof membrane of your RV/truck, you should choose the right type adhesive to use.

You can also use the solar brackets to mount the solar panels. Lay out the panels at the specific places you want to install them, then fit solar brackets under the panels. Under each bracket, apply a sealant and ensure it is water-tight so it allows no water leaks into your RV through drilled holes where the brackets were mounted. The type of sealant to use depends on the roof material of your RV. If you're not sure which sealant to use, you can confirm from your RV manufacturer or dealer. A Dicor lap sealant can work on almost all RV roof material.

Once all the brackets are in place, you can screw the panel into the RV roof securely. Solar brackets come in two forms: tilting and non-tilting. You can choose any of the brackets based on your roof type. If you have a flat surface and want to have a flat mount, then non-tilting brackets should be enough for you. Note that when screwing the panel, you should do so at the frames not to the panel. Accidentally screwing the panel tray is a costly mistake.

Make sure the solar panels remain covered with a cloth, box, or even cardboard when installing until you're ready to activate the system to produce electricity.

### Roof Wiring of the Solar Panels

Each solar panel comes with two 10 AWG wires with an MC4 connector at the end. If you have a combiner box located inside the RV, you will have to extend your wires to reach the combiner box.

You can use the MC4 "Y" branch connector to combine the positive terminal wires from the two solar panels into a single wire, and the two wires from the negative terminals into a single wire. The MC4 "Y" connector combines the four wires from the two solar panels into a set of two wires. This reduces the number of wires that are running across the roof down to the RV through a refrigerator roof vent. Power cables should run from the solar panels to the combiner box, then into the RV and down to the solar charge controller.

*Ways to run the wires:*

- **Refrigerator vent**: This is the most commonly used strategy in which the wires would run through the refrigerator vent and into the RV. You don't need to have any extra holes to run the cables.
- **Drilling**: If you don't have a refrigerator vent or anything else that can allow you to pass the cables, you can drill a new hole for that purpose. The new hole should be drilled at least near the cabinet or next to the interior wall of the RV. This will enable you to hide the wires inside the RV. Use a sealant or waterproof entry port that can cover the drilled hole to prevent water leakage into your RV.
- **Plumbing pipe**: You can look into this option if your refrigerator vent is far away from the battery compartment. In this case, it will be good to run the wires through the plumbing pipe. Add a sealant to any hole drilled on the roof.

You can also use a 15A fuse on your MC4 connector; just make sure to use an appropriate fuse according to the cable size of the panels. The MC4 connector wires should be installed at the end of each of the four extension wires if your combine box is inside the RV. Then, install the ring terminals on the other end of the extended wires. The ring terminals are to be installed in order to attach the wires to the combiner box before connecting the extended solar panel wires with MC4 connectors. You can also add Dicor sealant to hold the wires in place. You need to have an MC4 Crimper tool to connect MC4 connectors to the end of the solar wires.

Once you have all the wires running on the roof to the combiner box, or through the refrigerator vent, you can tie the wires together temporarily using electrical tape. You can also add a layer of Dicor over the wires to firmly hold them in place on the roof. If you're using a combiner box inside the RV, you can place it next to the refrigerator inlet vent. In doing so, there will be no need for a waterproof box on the roof. You can use two power distribution blocks to connect wires from the positive terminal, along with another block to connect the wires from the negative terminal. If you have ground deploy panels, you can use another block to connect the wiring coming from the ground deploy and another block to connect to the solar controller.

## Wiring from the Combiner Box to the Solar Controller

You can connect an AWG welding cable to run from the combiner box into the solar controller. The size of the welding cable will depend on wire calculations you did in your planning stage. At the end of each cable where you put power distribution blocks, you can attach wire lugs to the cables. The wire lugs will ensure the cables are well secured.

### Connecting **Solar Charge Controller**

Solar controllers should be connected close to the batteries to reduce loss on the wires. Mount the charge controller on the wall and connect it to the power cables you passed through the refrigerator vent. You can go through the instruction manual that came with the solar kit in order to do this. Some solar kits have specific wiring procedures—for example, in some solar kits, you will have to connect the solar charge controller first before connecting with the wires from the panels. Other solar kits will require you to connect the charge controller to the panels first. Therefore, make sure you go through the instructions and follow the recommended wiring procedures.

Use a multimeter to determine the polarity of the connection cables. Your connection to the solar panels and batteries should be as per the solar kit instructions or manufacturer's recommendations.

If your RV battery is on the open, avoid installing the solar controller there; instead, you will have to install it inside. For instance, you can choose to mount it inside the baggage compartment and run the wires from the controller into the battery. The compartment wall will provide you with a secure point to mount the solar controller. If you're mounting the controller on the compartment wall, you will have to mount a piece of plywood and screw it securely on the aluminum material to form a flat mounting base for the solar controller. In another piece of plywood, you can mount the solar controller, power distribution blocks, and a circuit breaker. This forms part of the pre-wiring of all components in the baggage compartment.

The circuit breaker is used to isolate the solar controller. In this case, you can use a circuit breaker between the solar controller and the solar arrays. You can mount another circuit breaker between the solar controller and the batteries.

### Sub-Assembly of the Charge Controller

If you choose to mount the charge controller inside the baggage compartment, you will also have to install other electrical components that can isolate the solar controller. This is achieved through the use of circuit breakers and distribution blocks to distribute the charge controller output.

In order to have a more organized compartment, you have to implement sub-assembly with all the required components. That way, you can have a single mount with all the components. A single piece of plywood is big enough to accommodate a solar controller, a pair of distribution blocks with both positive and negative terminals, and two circuit breakers. All these components should be laid in such a manner that allows you to run a 6 AWG cable or AWG welding cables of any size. Then, screw all the components on the plywood. Use one circuit breaker to connect the positive wire from the solar array into the combiner box. This disconnects the solar array cables from connecting directly to the solar controller. The other circuit breaker is connected between the positive terminal of the solar controller and the power distribution block, which passes a charge to the battery and power distribution panel of RV. This disconnects the charge controller from the battery. Using the circuit breakers isolates the solar charge controller from the rest of the RV's solar power electrical system. The work of the power distribution blocks (both positive and negative) ensures the RV batteries are linked to the power distribution center and the solar controller. As a result, the solar controller can allow the 12V solar power system to run your RV electrical appliances while providing power to recharge the batteries.

Use 6 AWG cables or less to connect the circuit breaker with the positive input of the solar controller. Connect the positive output from the solar controller to the other circuit breaker, then connect this positive cable into the power distribution block. Connect the negative output cable to the negative terminal of the power distribution block.

**FINAL INSTALLATION of the Solar Charge Controller**

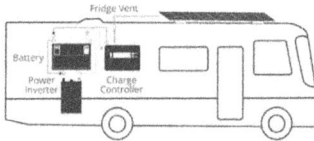

ONCE YOU COMPLETE the pre-wiring of sub-assembly, you can then attach it to the plywood wall you initially mounted on the baggage compartment. Once done, hook up the welding cables running from the solar panel via the combiner box into the power distribution panel and batteries. The positive wire terminal from the combiner box should go via the circuit breaker connected to the solar controller, whereas the negative terminal wire should be connected directly to the solar controller (the wires should be attached at the PV input side on the controller). Make sure to leave the circuit breaker in *open status* to ensure no power flows before you complete the installation process. That is, it should be left *open* until you **activate** the power system.

After installing the wires to and from the solar controller and the baggage compartment, you will have to fill up or seal the holes you have drilled. You can use a silicone sealant that ensures there is no dirt, moisture, or even bugs that can enter through the hole.

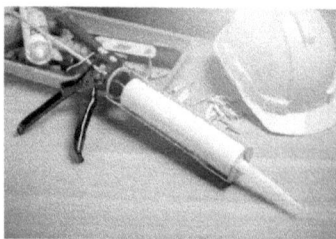

*Image source: Nor Gal/shutterstock*

### INSTALLING an Inverter (*optional*)

If you intend to use AC appliances such as a laptop, fridge, or microwave, then you will need to install an inverter that converts the DC from solar arrays to AC.

The inverter should be mounted near the battery and far away from any corrosive gases that may come from the battery or other damaging elements like heat. Always follow the instructions in the kit manual and ensure both positive and negative terminals are connected securely to the inverter. Connect the negative terminal first before connecting the positive terminal, then connect the inverter to your RV electrical system, ensuring the appropriate wire gauge is used for the connection.

### BATTERY INSTALLATION

Battery installation should be the final part of your solar power system installation.

Batteries are very important, especially when you are camping. You can choose a battery based on your solar power needs. Connecting the batteries to the charge controller ensures the battery is charged to full capacity while not overcharging,

*Image source: chonlawut/shutterstock*

TO CONNECT THE BATTERY, you can use jump wires

to connect to your RV (both ground wire and positive terminal wires). Make sure the wires are of an appropriate gauge; using smaller wires may heat up and explode.

The black wire would connect to the positive terminal of the battery—that is, the positive cable would run from the distribution block in the solar controller up to the battery through the use of a catastrophic fuse. The ground wire or the negative cable should be hooked to the chassis, then to the frame of your RV or trailer. The frame acts as the path to the negative or ground terminal for your 12V electrical system. The catastrophic fuse would connect directly to the positive terminal of the battery.

### Connection of 12V Dual Batteries

Just like connecting a single battery, if you have two batteries, you can decide to connect them in parallel or series and increase your camping power. Connecting the batteries in series increases the voltage, but it doesn't increase the amount of time required to recharge the batteries. Before installing, you will need to figure out the positive and negative terminal in the battery.

A parallel connection connects all positive terminals together, along with negative terminals together. When using a 12V battery, the voltage will remain the same, with the battery providing electricity for a longer time. In a series connection, the batteries will add the voltages together when a specific current is attained. While in a parallel connection, the batteries add a current that has the same voltage because the lead acid is always affected by the current that is drawn from the battery, not its voltage.

The positive wires are black in color, whereas the ground wire is red. To connect the battery in parallel, install them on the frame of the camper. The camper's positive cable should be hooked to the positive terminal on the battery, whereas the ground wire from the frame should be connected to the negative terminal on the opposite battery. This ensures the batteries' drain charge at the same rate. Connecting both cables to one battery will make one battery drain faster than the other; a dual battery connection extends the battery's life and improves their efficiencies. In another case, you should connect batteries in parallel if you purchased the two batteries at the same time; otherwise, if you connect an old battery to a new battery in parallel, you will reduce the efficiency of the new battery and its lifespan.

Take another jumper wire and connect from the positive terminal in one battery to the positive terminal of the other. The jumper cable should have the same or greater gauge as that of your camper cable. Make sure the terminals are clean on both batteries. Hook another cable from the negative terminal to the other negative terminal on the other battery. Now, your battery will be able to supply you with 12V power for your RV appliances.

### INSTALLATION OF BATTERY Monitor

Battery monitor is a very important part in battery installation, as it monitors the flow of current in your batteries.

The monitor consists of a:

- Shunt
- Display

The shunt monitors how the current flows in and out of the battery. It connects to the battery load to read the load and shows the battery's charging and discharging rate on the monitor display. If the load from the battery passes between the shunt and the ground post of the battery, then the shunt can't record the load because it simply can't see it and doesn't know whether that load is using battery power. To avoid this outcome, you will need to relocate the ground chassis cable and connect it somewhere where it would not be after the shunt.

. . .

### Mounting the Shunt

Mount the shunt close to the battery to ensure there are no loads between the shunt and the ground/negative terminal of the battery. Ensure your battery and the shunt are not in an open area where they are exposed to weather elements. Even when you use battery boxes, water can still manage to get inside, which you should avoid as much as possible. Therefore, it is best to install the battery monitor in an area where the shunt can't get wet.

If you don't have enough space to install the battery and the monitor inside, you can buy a plastic weatherproof junction box and install it upside down against your RV or trailer floor. Start by screwing the monitor shunt into the junction box. You will also need to drill holes and pass the ground wire to connect to the shunt. In this case, the shunt and the monitor display will communicate with each other via a cable. The positive cable will connect the positive terminal of the battery and the shunt, which ensures the shunt receives the power and reads the voltage coming from the battery. After installing the shunt in the junction box and have tested that it's working properly, seal the drilled holes using a sealant. For this step, you can use a silicone sealant to ensure no water or moisture gets inside the junction box.

. . .

### *Mounting the Display*

You can mount the monitor display anywhere in your RV—that choice is yours. You can also have a long cable that connects the display and the shunt. Do your best to install the display where you can constantly monitor the flow of current in and out of the battery. You can install the monitor display in your living area or on your pantry to monitor the battery while sitting in the booth in your RV. Once you find the suitable location for your monitor display, then mounting it becomes very simple, as you would only have to secure the mounting lock ring on the identified location and plug in the cable coming from the shunt. You can pass the cable used for communication between the shunt and the display in the same place you ran the cables connecting the combiner box and the solar controller.

### Tips to Extend the Battery's Lifespan

- Ensure the battery drains to approximately 50% before charging it again.
- Never store a discharged battery because doing so will crystallize and harden the battery plates. Hardened plates can't generate power.

- The RV battery discharges when not in use; therefore, you should recharge the battery after every three months.

## Caution While Charging the Battery

- Ensure the surrounding battery area is well ventilated when charging.
- Ensure you have corrosive-free and secure battery terminal connectors.
- Keep the charger away from the battery, as dictated by DC cables.
- The charger shouldn't be mounted above the charging battery. Gases emitted can corrode and damage the battery, or even cause an explosion.
- Never charge a battery that has already frozen. If the battery temperatures are lower than 0°C (32°F), then warm the battery until it is at room temperature, then charge it.

## Measuring Battery Polarity

Generally, a 12V battery would be connected using a black cable for a positive connection and a white cable for a negative or ground connection. Sometimes, you can wire the rig with a red cable for positive connection and black cable for ground or negative connection.

If you reverse the battery polarity, it will damage all the 12 DC appliances connected to your RV; therefore, before disconnecting the battery, make sure to mark the positive cable, and your work will be easier.

### Activate Your Solar System

Once you have installed all the components in place, then you will have to activate your solar power system to enjoy its benefits. Connect the positive lead cables to the battery, then the negative lead to apply power to the entire electrical system. If your 12V system is wired correctly, the RV should turn back to life!

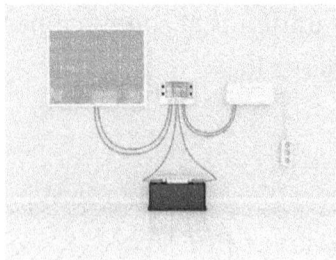

*Image source: styleuneed.de/shutterstock*

**Turning on the Other Solar Components**

Turn on the circuit breaker between the battery and solar controller to activate the solar controller. This will enable you to inspect the solar controller before activating the solar arrays. Ensure the solar controller settings are correct (look at the app settings and the rotary switch). Also, check whether the LEDs blinked. After you're satisfied with solar controller settings, climb to the RV roof and remove the cardboard or cloth you used to shade the panels from charging.

Next, your solar arrays should generate power if it's a sunny day. You can verify the power being generated by looking at the app settings on the solar controller. You will be able to see how much power is being generated by solar arrays and how much is being sent to the batteries. You can also confirm the flow of electricity with the battery's monitor.

**Chapter Summary**

Installing your solar power system is very easy, and you don't have to be an expert to install the solar system in your RV, boat, or car. Before beginning your installation journey, you will need a plan for all your solar needs. Solar system planning enables you to know how many solar panels you need to buy, the battery size, the wiring required, and the size of all the other solar components you will need.

Once you're able to figure out your requirements, the next step was to learn:

- How to install the solar panels.
- Wiring of the solar panels.
- Mounting the combiner box.
- Installing solar charge controllers.
- How to install the batteries.
- How to install the batteries' monitors.
- How battery monitors work.
- Activation of the solar system.

In the next chapter, you will learn how to install 12V DC and AC appliances.

## HOW TO INSTALL SOLAR PANELS IN YOUR CAR, VAN AND BOATS

⚜

etting Up a Solar Panel in Your Van/Car
Assume you have two 100W solar panels and you have to mount them securely on your van's roof rack. The panels will generate electricity by harvesting solar energy from the sun, and the energy generated will be passed via a series of adapter cables up to the solar charge controller. From the controller, the generated solar energy flows into the batteries through a series of other cables.

This energy is stored in the battery, awaiting to be used. You can run cables from the battery to the inverter, which changes the DC to AC. From the inverter, you can connect any other device you want to charge or power.

.  .  .

## Mounting Solar Panels on the Roof of Your Van/Car

One of the most difficult tasks that people often face is how to mount the solar power system on the roof of their car, van, or truck. You need to figure out how to mount the panels securely on the roof of your car or van. The panels should be mounted in such a way that they won't fly off while driving at high speeds and are also protected from other elements and aerodynamic motions.

### How to Mount the Panels

**Step 1:** Mount the solar panels onto a thin plywood board, or use a sheet of metal to mount them. You can get the plywood from your local retail hardware store.

Make sure you buy plywood of the right size (it should be able to fit the shape of your 100W panel). You can also buy a plastic sheeting and secure the plastic sheet on the wooden board. You can use glue to fix the plastic material firmly onto the board.

**Step 2:** Once the plastic sheeting has dried out around the board, you can then mount the panels on the top of the board. Drill some holes at the edges of the panels into the board below. Then, secure the panel in place by using bolts and nuts.

Each solar panel should be secured with six bolts/nuts.

Some solar panels come drilled on the sides, so you won't have to drill them—you will only have to fix the bolts or nuts to secure them firmly on the board. Once the solar panels are mounted on a flat and waterproof board, the next step would be to mount them on the roof of the van or car.

**STEP 3:** Our next step would be to mount the entire solar array to the roof of the car/van. To mount on the roof of the car, van, or truck, you will need to have a roof rack. Drill two or more holes on the roof rack to fasten your solar arrays. Some roof racks have at least four holes to install the nuts, whereas others will have 6 to 8 holes that will allow you to install components on the rooftop.

Position your solar array board in the exact location you have attachment points in the roof rack of the car or van. That is, the board should line up with the attachment points, mark position of these points on your solar array. Then, drill a hole through the board that will be used to secure the solar array board on the roof rack of the car. Do not drill holes on the solar panels. Take a bolt that can be screwed through the board up to the roof rack. Always make sure to use the appropriate material and size based on the material used to make the roof rack. Screw the bolt into the attachment point of the car roof rack, then equip it with a washer. The wider the washer, the better.

STEP 4: Position the bolt and washer on top of the solar array board on the vehicle's roof rack. Screw the bolt into position and repeat the same procedure to all the other attachment points on the roof of the car/van.

This installation process will depend on the model of the car/van you have, the type of the roof rack, along with other factors; therefore, the installation may be different for different people, but you should be able to mount the solar panels on the board. Later they should be fastened to the roof rack using bolts and washers screwed along the attachment points. Once you're done with mounting the solar panels on the roof of the car, cover the panels with a heavy blanket or cardboard to avoid charging them, then proceed with the next step of installation.

**Note:** The solar arrays are covered so you won't burn the solar charge controller or other devices when you connect the wiring from the solar panels to the solar charge controller. The panels should only be uncovered once the complete installation is done and all other electronic components are in place and working properly.

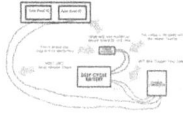

**STEP 5:** Where you would install the battery, charge controller, and power inverter depends on the model or type of vehicle you have. For example, if you have a van, you can have the battery stored in the floor of the center compartment, while the charge controller and inverter can be stored under the front passenger seat. You can mount the components differently, as there is really no limitation for where to install them. Always make sure the cables connected to the solar controller, battery, and inverter are fastened well, such that they don't move or come out when driving the vehicle.

Once all the components are connected with the right wire size, you can then activate your solar power system. Remove the heavy blanket covering the panels on the rooftop of the vehicle. You should now start to see some actions recorded by the solar controller display unit. The angle and intensity of the sun determines how much you can generate in amps.

. . .

## Installation of the Solar Power System in a Boat

Solar panels provide you with a reliable and cost-effective way to maintain your boat with low fuel charges. The solar panels on a boat work like other off-grid solar power systems.

The solar panels generate electricity needed to support the boat power needs. The solar charge controller controls charge input to the battery to avoid overcharging or damaging it, while ensuring the battery doesn't receive more voltage than it can handle. Most electronics in your boat should be running on DC electricity, though other electronics like TV and microwaves need AC electricity; therefore, you need inverters to convert DC to AC electricity. Some complete solar panel kits exist for marine systems including wires, cables, and mounting equipment. You can also get these components separately.

Before adding the solar panels to the boat, you will need to ask yourself if you need the solar power to power the entire boat or just the electronics inside. If you need a solar power system for the whole boat, you will have to invest heavily on the solar system. Determine the daily power needs for each electronic device you use in your boat—do you live in the boat full time or do you only use it on weekends? These are some of the factors you will have to consider when calculating the power needs of your boat. Finding the total daily consumption will help determine the size of solar panels you should need, as well as the battery size.

### Finding Space for the Solar Panels

One of the challenges many people face is finding enough space to install their solar panels. Though modern designs of these panels allow them to be of a smaller size; however, they will still take a lot of space. Therefore, before buying the panels, measure the layout and dimension of the boat. If you have a small boat, you may not have enough space to store the panels. Some sailors mount the panels on the sides of the boat by embedding them on the deck, whereas others have crafted overhead structures or racks to mount their solar panels. You can also purchase foldable solar panels that you can take out whenever needed.

### Avoid Shadows

When mounting the panels either on the sides of the boat, on racks, or using fordable panels, ensure they're not blocked by anything. If the sun's rays are blocked from hitting the panel surface, such will reduce the energy produced.

Note that solar panels on a boat are always

affected by cloudy skies, as the panels get hit by energy-dampening shadows that change based on the direction the boat is heading and the location. When installing the panels, make sure they're far away from any drastic shadows. You can also decide to have more solar panels to offset any unavoidable shadowing.

## Mounting the Panels

After figuring out the size of the panels you need, the next step is to determine the location to mount them. The panels should be located where there are no shadows and also in a place where they can't interfere with the boat's operations. They should be installed in such a way that allows you to rotate them toward the sun as you move around on your boat. This will help you increase power generation by 40%. Some ideal places you can mount the boat include the cabin tops, radar arches, stern rails, Bimini tops, and at the top of dinghy davits. If you have rollable or marine-grade solar panels, you can mount them in different areas using the corner grommets. These panels are easy to mount, remove, and store. Mount the panels on a raised surface to allow the air circulate beneath it. High temperatures beneath the panels can increase resistance and reduce the solar cells' outputs.

. . .

### How to Mount the Panels

There are various methods you can use to mount the panels on the boat, which will depend on the type of the boat you have and its roof surface.

Some of these methods include:

**Use of glue adhesives**

If you choose to use the glue method, you will have to do some preparation prior to the installation of the panel. Cleaning the surface is highly recommended before applying the adhesive. Some manufacturers advise to lightly sand the surfaces before application.

### Use of bolts

Depending on the type of boat you have and where you use it, if you have a yacht cruising the sea, then glue adhesive would not be ideal for you. To hold the panels securely on your boat's roof, you will need to drill holes on the roof's surface and fix the panels with bolts. After fixing the screws on the holes drilled, you can use a marine-grade sealant to fill up the holes and avoid any leakage into the boat.

### Panel size vs. type of mounts

Panel mounts are not ideal for different-sized panels. Fixed mounts are the most robust and will be good for those who cruise a lot. The mounts allow you to fix the panels on a flat surface permanently and can handle any weight.

If you want the solar panels to be tilted at an angle, then you need to confirm with the manufacturer about how much weight the mount can handle. 100W panels can be accommodated in cheaper mounts, and if you have more than 100W panels, you should confirm the maximum load the mount can hold.

### WAYS TO POSITION the Solar Panel Mounts

You need to find the right position for your panels on the boat; otherwise, the wrong position will affect the performance of the panels. For example, if you use a fixed mount on the panels on a deck, then you can use a rectangle canopy to shield yourself from the sun if it blocks the panels. In this case, no energy will be produced, resulting in a waste of your money.

The mount type is very important, especially if you're stationary for a long period. Tracking the sun so you can tilt the panels will help increase the amount of power you generate. If you're moving constantly, mounting flat panels will help you maximize the sun's rays to generate more power. Once you identify the right position for the panels, the next step will be to come up with a layout for the exact position to mount them. Drill some holes where to fix the panels—the drilled holes should match the ones in the panel frame. Fix the panels securely using bolts or t-nuts. Once mounted, cover the panels. The solar controllers and the battery should be installed inside the boat, and the cables from the solar panels should be connected to the controller that is attached to the boat.

## CHAPTER SUMMARY

Installing solar panels in your land vehicle or boat is very easy. All you need to do before installation is figure out the size of your solar panels and have a layout and wiring system to install a complete solar power system.

To figure out the number of solar panels to install, you would have to calculate your daily power consumption from all the smart equipment connected to car, van, or boat. Knowing how much power you need per hour to power a particular device will help you figure out which solar panels can produce an output equal to or are more than your total power consumption from all connected devices.

The layout of the panels is very important, especially when installing on a boat's rooftop. The panels should be away from shade and mounted securely on the rooftop.

## ADDING 12V APPLIANCES

~~~

*A*dding DC Appliances

A variety of low-voltage DC systems are built to run on trucks, cars, RVs, and vans. Using DC to run various applications makes off-grid solar power cost much less. Modern devices operate using DC without the need to convert them to use AC, which will cause double the energy conversion. Using both DC and AC provides you with a better working solar power system. DC load appliances are significantly cheaper and they consume little amount of power, making them the most preferred for a sustainable solar system.

DC is well-suited for small appliances that require less power. Some of these appliances include: cellphones, LED lights, alarm system, fans, water pumps, refrigerators, laptops, among other similar items. These appliances are built with 12V power, while other home appliances are available at 24V power. A voltage converter can increase your battery voltage from 12V to 24V or 48V. Devices like cordless phones, motion sensing light control, alarm systems, and doorbells can work even when the inverter is switched off; therefore, you shouldn't have to hesitate in using DC to power some of the appliances in your vehicle.

### How to Add Interior Light to Your RV

Light emitting diode (LED) is a semiconductor component that emits light once current flows through it, which only flows in one direction. LED lights are mostly preferred over traditional bulbs that waste a lot of energy as heat. LED lights use less power and are both more durable and longer lasting.

LEDs use the principle of semiconductor junction, in which different materials join together to form a diode. The diodes allow electricity to flow in one way while also producing light. It uses a positive lead (anode) to connect to the positive terminal of the power supply, whereas the negative lead (cathode) connects to the ground wire.

### How LEDs Produce Light

When diodes pass electricity through the semiconductor material, they produce light. The light's color depends on the material used. LEDs operate on low voltage of about 1.5V, unlike regular light bulbs. As a result, resistors are used to limit the amount of current flowing. When you power the LED lights via your car, a resistor is used to avoid burnout of the bulbs.

## How to Install **LED Lights**

Installing LED lights is very simple. In just a few steps, you can install lights to your vehicle. You can easily achieve that through proper planning of the required components. The next steps you would take are to:

- Wire the lights directly to the power source on your 12V battery.
- Mount the light strips securely.
- Test and verify if the lights are running.

LED lights don't need a lot of wiring; therefore, there will be no need to run wiring to the battery. The lights will also require little power, hence why you can connect them to the factory stereo of your car or use the cigarette lighter socket wiring.

You can wire the 12V LED lights to the cigarette lighter plug socket wiring, which should have an on and off switch. You can also use a fuse box in the vehicle's interior or even behind the car stereo. The set is hardwired to the accessory wire to obtain a +12V power supply, which will enable it to switch on and off when the ignition is turned on.

### Finding the +12V Accessory Wire
#### Check your vehicle's wiring colors

You can check your car or truck's color wiring codes from 12volt.com. The website displays diagrams with colors for each type of car, making it easy to find color codes for your vehicle.

### Test the wiring

You will need to use a multimeter to test the color code wiring to find a suitable wire. The reason you have to test it is because some modern vehicles don't use 12V wiring, but instead a wiring voltage of below +12V.

If you have a radio in your car, remove it and put the ignition in accessory position. Look for +12V wiring from the set. Switch off the ignition and test again to determine which of the wires is suitable.

### Use a fuse box

A fuse box houses the power supply for the LED set and radio, and it can also be found on the driver's side next to the brake pedal or under the panel in a dash. You can connect the wires to the power source in a fuse box. You can find them on the left side of the dashboard between the lower left side of the car interior or under the panels. The car's manual should also list all the fuses, along with the function of each fuse.

Fusebox powering adapters make it easy to tap into the power circuit power. When tapping off on the fuse box, you can use adapters to make your work easier. If there is no manual for the car, you can use the test meter to determine the power source supply.

### When to Turn on the Lights

Use under dash lights with different settings when you turn them on. They can be on when headlights or parking lights are on or when you start the vehicle. The lights turn off once the car shuts down. The fuse that comes with the headlights are sometimes difficult to trace. You will have to find the fuse since it controls the interior lights of your vehicle. The lights can be labelled as the interior lamps, dash, lights, etc. In those that came with the vehicle, there is more than one fuse connected. In this case, it becomes more straightforward to use. One of the available options is to put the lights on the same fuse, like the radios, thus when the radio is on, the lights are on too. The fuse can also be used for ignition, such that whenever you turn the ignition key on and the car is running, it will turn on the lights.

You can use any of the above methods to turn on the lights. Use the test meter to determine the power supply of the fuse, which can be done by connecting the black probe of the multimeter to the ground port of the vehicle, with the red probe being connected to the fuse. If no voltage is detected after testing, then test the other side. If no voltage is detected after that, then slice into this side. Doing so allows you to locate the back side of the fuse box and the cables that are connected to the cables on the fuse side. Slice it in line with the cables using connectors.

### Identify the Grounding Point

This is a very important step in the process. Identify an area under your car or RV dash to use as the ground. The identified area acts as the firewall, and it separates the car engine from the inside of the vehicle. Locate spots on the firewall that are bolted, then remove one of the bolts and use connectors to slide under the bolt. This forms the ground wiring.

### Use an Inline Fuse

Use the red cable from the tester to connect to the fuse. Adding an extra inline fuse to the connection circuits increases the safety. Connect the inline fuse close to the connected fuse box.

### CONNECTING **Wires to LED Strips**

The cables connected to the fuse box and those connected to the ground should be connected to one of the two LED strips. Slice these cables together and add another cable to connect to the second LED strip. After connecting the cables, run them under the dash and connect them to the strip.

### TESTING **the Wires**

You have to test the connected cables to ensure they're working properly. The testing procedure depends on the wiring system of the connected wires. Whatever wiring is used, try switching on the lights. If the lights do not work, you can troubleshoot it to try to identify the problem.

### Mount the Strips

It's now time to mount the strips in the interior of the car. The plastic material under the dash is an ideal place to mount the LED strips. The wiring should be kept out of the way of all moving parts. Remove the strips of plastic material covering the adhesive, then stick it to the surface.

. . .

## Tidying the Wires

All the wires connected to the strips should be tied up using zip ties to keep them out of sight. The best way to do this is to tie the wires to any existing wiring harness in the vehicle. Almost all vehicles have a piece of harness wiring that runs under the vehicle dash.

## Troubleshooting

If the lights don't turn on after the wiring, then you have to find where the problem is. This can be done by:

- Checking whether the fuse is well connected.
- Checking if all connections are fixed securely.

If only one light is working, then check whether the wires are well fixed to the strip and whether the LED strip is ready for hot wiring directly from the battery.

## Powering a Laptop Without an Inverter

If you love using your laptop while camping to access the internet or upload pictures, you will need to have a 12V laptop car charger—not necessarily an inverter. A reliable laptop charger will ensure your laptop is charged at all times while, at the same time, not draining your entire battery.

Connecting the laptop through the invert is not energy efficient, so you can use a charger to connect the laptop directly to your 12V system through the cigarette lighter socket.

Requirements to connect to 12V system:

- 12V laptop charger for your laptop.
- Plugs that enable you to connect the laptop directly to the cigarette socket.
- More efficient energy.

The quality of DC energy supplied to your RV or car will be powerful enough to power your laptop without interfering with the running of other appliances in your car, like the audio equipment.

You can buy a universal 12V laptop charger with a power output of 90W which is enough for charging most laptops. The charger that plugs into your cigarette lighter socket will also have USB ports that allow you to connect to either your phone or tablet, which is a convenient and energy-efficient way to connect your laptop to your batteries without the use of inverters that can drain the battery.

. . .

## OFF-GRID INTERNET CONNECTION

Internet connection is considered part of daily lifestyle in the modern era. When camping, you may go to locations with little or no phone signal, cable, or network connection, thus the lack of internet becomes a crucial concern. However, thanks to emerging technologies, satellite Broadband enables access to the internet anywhere, no matter how remote the place is. It ensures you remain connected whenever you're going off-grid. If you love the off-grid lifestyle, you don't have to worry about how you plan to stay connected. The modern satellite broadband ensures you stay connected wherever you go. With the internet connection while boon-docking, you can run an online business, work away from the office, stay connected with family, and much more.

### Satellite Internet

To stay connected wherever you travel, you need to have a satellite dish or antenna and transmitter connected to your RV. The antenna should be mounted on the roof with a clear view of the sky, which can ensure you have a stable and fast off-grid internet connection.

Though satellite internet is very expensive when compared to cable internet, it provides you with independent, renewable, and remote off-grid connection. Satellite internet uses a speed of 22 Mbps, which is similar to the speed of standard ADSL internet. HughesNet and WildBlue are some of the best companies that provide affordable satellite internet almost everywhere.

If you're watching the power consumption in your RV, you can shut down the satellite modems when you're not using them, since they can consume 20W to 30W of power.

RV satellite systems can be very expensive. If you're looking for a roof-mounted system that allows you to have access to the satellite signals wherever you go, be prepared to part with a couple thousands of dollars for buying the hardware equipment needed, along with the set-up costs.

### Pros

1. Access to the internet in areas where cell towers are unavailable.
2. Access to fast download speeds.

### Cons

1. During peak hours, you may experience a reduced download speed.

2. The cost of the hardware required to mount the system on the roof is very high.
3. Bad weather in the area in which you're camping may affect your signals; you need to have a clear view of the sky.
4. When using certain services, such as Skype, you may experience latency issues.

## OTHER OFF-GRID INTERNET OPTIONS

### 1. **Cellular phone connections**

There are various remote areas where you can have access to cell towers. If you're camping within the range of cell phone towers, you can use your data-enabled device to access internet facilities and surf the web.

Cell phone internet connection is one of the cheapest methods on the market, which also means that you may sometimes experience slow connection. You can use cell phone internet in times of emergencies, especially if the main internet service is down. It allows you to tether the phone's internet for use in your laptop to accomplish your internet goals. Cellular internet connection is ideal for those who don't need a lot of bandwidth.

.   .   .

## *Pros*

1. Provides you with the cheapest off-grid internet access options.
2. Mobile internet access from anywhere and at any time.
3. Allows you create a portable Wi-Fi hotspot or tether your phone's internet to be used on other electronic devices.
4. Allows you to make and receive calls through a dedicated phone line.

## *Cons*

1. Slow connection speed compared to other forms of accessing the internet.
2. Has a small screen size access.
3. Not efficient in downloading large files or watching videos.

## Using wireless internet access

If you're within a particular mobile broadband repeater coverage, you may decide to use a wireless turbo hub or stick to obtain off-grid internet.

Some portable internet hubs come with backup

batteries to help you prolong the time you take to access the internet. If you want to use the available wireless option hubs, you can buy a rocket stick, which is cheaper and plugs into the USB port of your laptop. Note that before you invest in an internet hub or rocket stick, find out if you can get the service in the area you will be in. You can talk to people or businesses around that area or get more information from the service provider. You can also buy a 3G or 4G hotspot device if you're within coverage. Some devices you can use to create your own Wi-Fi hotspot include the Clear Spot Voyager Wireless Hotspot and T-Mobile Sonic 2.0, both of which can provide you with a 4G hotspot.

The hotspot devices come with an external Wi-Fi antenna and a signal amplifier that can help you increase your range of coverage. This one of the more reliable off-grid internet options if you want to stay connected even in the remote areas.

### *Pros*

1. Access internet connection easily through multiple devices.
2. Provides you with a fast connection speed compared to cell phone internet.
3. Cheaper compared to satellite internet cost.

## Cons

1. Expensive compared to cellular networks.
2. There is an extra hardware cost to enable you access the Wi-Fi hotspot.

CHAPTER SUMMARY

In this chapter, you learned how to add various 12V DC appliances to your car or RV. Modern devices are built to use DC, and there is no need to convert them to use AC. There are various DC appliances you can install in your RV system that use a 12V system. These devices include: LED lights, air fans, water pumps, cellphones, and laptops, among others. Other home appliances can support a 24V system, like refrigerators, the alarm system, doorbells, motion sensing light control, etc.

Voltage converters increase the voltage of your system from 12V system to 24V, 48V, and other voltages based on your battery size. Therefore, don't fear adding your favourite electronic device in your RV system. Small DC systems consume less power, which makes them the most preferred when dealing with solar power energy.

You also learned how to install a laptop to your

solar system without the use of inverters. Another DC appliance you can install to use the battery energy is interior LED lights. You can also install AC appliances into your RV, or car. Installing inverters to the system will allow you to convert the DC into D. In this chapter, you learned how to add AC appliances to your RV, like with the use of off-grid internet services. You can rely on satellite internet when camping in remote areas or have access to other internet options.

In the next chapter, you will learn about the operation of various solar panel components, along with the maintenance of those components.

# OPERATION AND MAINTENANCE OF RV SOLAR SYSTEMS

$\mathcal{CW}$iring and design of the solar system are some of the more overlooked areas in PV system design. Careful design of the solar power system and proper wiring will result in more efficient and reliable solar energy production. Safe practices and adhering to proper electrical standards and code will result in reduced hazards caused by electrical appliances. Therefore, proper maintenance is essential to extend the lifespan of your solar system.

## SYSTEM PROTECTION and Safety Considerations
### Cables and Wiring

If the cables are connected improperly, it can sabotage the functioning of the solar system design. Choosing the wrong wire size affects the rate at which the battery is charged and can sometimes cause system failure.

## *Cables Considerations*

- Use array cables, as they're more suitable for DC current.
- Use UV-resistant string cables.
- Select cables based on IEC 60548 standard: the design requirement for installing PV solar panels.
- Use double-insulated cables, especially when they're to be laid in a metallic tray.
- The total voltage drop should be less than 3% in DC cables, while on AC should be less than 2%.
- Install the cables in such a manner that all the wires are protected from any mechanical damage.

## Charging Battery with DC Power

There are various ways to charge your battery without having to pass through the converter first. The DC power makes it hard to run applications using AC. Converters help charge the battery while sending power through your inverter to enable you to power AC appliances. This will allow the battery to power both AC and DC appliances at the same time. Using inverters to change the DC to AC current through your battery reduces the efficiency of the appliances compared to when powering them directly via solar power. Charging your battery while running other appliances makes your battery take a long time to fill. You can also charge your 12V battery using your car's alternator. This happens whenever you restart the vehicle, as it will charge your battery as you drive. Using an alternator to charge the batteries enables you to have an extended road trip.

### How to Replace Your Battery

1. Switch off all the power, such as lights and other appliances. If there is a spark from the negative cable, it is an indication of a connected power draw device.
2. Note down the battery position in terms of its polarity.
3. Disconnect the cables starting with the negative cable, then remove your old battery.
4. The hold-down hardware and battery carrier should be free of any corrosion.
5. If the new battery is not maintenance-free,

you can top it up with electrolytes—that is, add distilled water to the battery until you achieve the right volume. During electrolysis, only the water level drops in the battery, though the acid remains the same.

6. Keep the metal at the terminal post shining. Use a wire brush to clean the terminal posts.

7. Install the new battery in the same polarity position as noted down in step 2 above.

8. Connect all the disconnected cables to the new battery.

9. Install the hold-down hardware.

10. Test the battery by switching on the RV lights.

### Temperature Regulation

Learning how to regulate temperatures in your RV is a great way to maintain a better life for your RV in great shape. You need to know how to run the heat pump and propane furnaces properly and more efficiently, run both AC and DC units, and control the remote temperature sensors in your RV. If you don't understand how the heating system works, you may end up spending more money on propane.

Regulation of temperatures inside your RV limits the amount of propane needed in your RV.

### Thermostats and Temperature **Sensors**

Modern RVs have a thermostat preset to operate under specific climate control. Most RVs have two thermostats, namely zone 1 for controlling the living area and zone 2 to control the RV bedroom. Some RVs have remote sensors attached to them, which is different to home systems that have the sensors attached to the thermostat. If your RV has the temperature sensors and thermostats installed, you need to know the exact location of each of the thermostat. This will enable you to conserve propane much easier during the winter, along with electricity generated during the summer.

### How to Adjust **Temperatures**

The majority of off-grid solar owners are looking for ways to adjust gas refrigerators to proper temperatures compared to those who use electric refrigerators. RVs, cabins, and vans do not have automatic climate control, hence the need to come up with ways to regulate temperatures during the winter and summer months. Off-grid refrigerators require manual temperature regulations since no digital thermostats are installed in the cabin or your RV.

Campers and those who love weekend getaways prefer using propane gas refrigerators, which are usually turned on during camping days and off when they leave the camping site. The refrigerator can stay for about 10 to 12 hours in temperatures between 70-80°F (21-27°C) to cool down completely.

To continue using the refrigerator during this period, you need to:

- Keep its door closed as often as possible.
- Set the refrigerator temperature to maximum.
- Don't add any food or liquid to the fridge if it is not frozen or cold.

After the cool down period, you can then start adding more food to your refrigerator in smaller quantities. If you load more food at ambient temperatures, your fridge may take a longer time before it can recover to safe food keeping. However, note that adding frozen or cold foods has no effect.

Loading the fridge with liquids at ambient temperature will take a lot of time and energy to cool. Using smaller containers to store liquids increases the air contact surface area; as a result, the liquids will cool more quickly. You should always keep the thermostat at maximum temperature and also replenish the ice trays every night. Putting a lot of drinks in a fridge will strain your fridge in terms of gas absorption and electric consumption.

### RV Heating and Cooling

If you love dry camping or boondocking, you may not enjoy it as much during hot days. However, you can get more creative and cool your RV without turning on the roof A/C unit. Although the A/C may be the best solution, it's very energy consuming. There are various ways you can cool your RV during the summer months without overworking your batteries.

You can start by increasing ventilation in your RV. Close all the vents in your RV bedroom and open the vents in the living area together with front A/C. You should also close the back door of your RV and crank up your A/C units. This ensures cold air is confined within your sitting area, allowing your room to cool down quickly. Having a ceiling vent can also help move air inside. You can also decide to park your RV in a shady tree to reduce the sun impact as well. Cover the windows of your RV with a reflective bubble material to reduce the amount of heat transfer.

Refrigerators generate a lot of heat when cooling food. When this heat is trapped in your RV, it will result in excessive heating. You can reduce the amount of heating coming from the refrigerator by cleaning the refrigerator vent to ensure no debris blocking. You can also install an air fan to reduce heat generated in the ventilation area.

Switch to LED lights to reduce the amount of heat generated by the halogen lamps that probably came with your RV. If you have shower skylight, you can replace the inner cover with a reflectix insulation, since a large amount of heat would enter your RV through this exposed area.

When driving in extreme heat, especially during summer, you will be forced to turn on your air conditioner because the built-in HVAC may not be enough to cool down the RV. The heat generated from the sun will be enough to power your roof A/C.

## Heating Your Vehicle

Most RVs have a furnace to heat the vehicle because they're easy to use, safe, and warm the whole rig. The built-in heater is inefficient and may take up valuable space. There are various alternatives you can use to heat your RV, however. Some heating equipment that can supplement the furnace include:

## Heat pumps

You can install a heat pump in your RV to warm your RV during the cold months. Some RVs have a built-in heat pump that uses AC to warm your room.

Tips to know when using heat pump:

- If the temperatures are below 45°F (7°C), the heat pumps will not function.
- The heat pump should be plugged into a plug of 30A and above—never run it under a 20A plug.
- Cranking the thermostat when using the heat pump affects the propane furnace. This may result in wastage of propane gas and electricity; therefore, when heating your RV with a heat pump, you should increase the thermostat temperature to prevent the propane furnace from kicking in.

## CATALYTIC PROPANE HEATER

The catalytic propane heater warms your rig without producing any harmful gases that can be generated by other propane heaters. This type of heater is portable, more efficient, and provides you with a concentrated warmth to keep you warm day and night.

. . .

### Ceramic heaters

This is a portable electric heater that provides warmth to a small area using minimal solar power. The heater helps conserve your batteries and are safer to use.

### Radiant Electrical Heating

RV owners are now beginning to embrace a technology only reserved for high-tech homes. Radiant electrical heating is installed under the floor and warms your RV from below. Though this electrical heating is not as efficient as other heaters, it will still make you feel more comfortable.

### Insulating the window

Windows also allow a lot of heat to escape through them, so insulating your windows with heavy binds and curtains will reduce the amount of heat that escapes through the windows.

### Chapter Summary

Maintenance of your solar system is very important. You need to learn safety and protection measures when handling solar power systems, safety precautions you have to follow when wiring the solar system, and types of the wires to use in connecting different solar power components.

Temperature regulation of your solar power system is also very important. Learning how to adjust thermostats and temperature sensors will save you a lot of money in the long run. Control heating and cooling of your RV so you can have a comfortable life and enjoy your camping and boondocking during the hot months of summer and cold months of winter. Learning how to control heating without using an AC unit, which consumes a lot of energy, is essential.

In the next chapter, you will learn some of the top mistakes to avoid when installing solar power systems for your cars, boats, and RVs.

# 12 MISTAKES TO AVOID WHEN INSTALLING SOLAR POWER SYSTEM IN YOUR RV

$\mathcal{A}$s the need to use clean, reliable, and renewable sources of energy, many people are switching to using solar power energy for their homes, cars, RVs, boats, vans, and even in their trailers. Many people have also switched to green sources of energy as a way to reduce electricity bills, which may result in a considerable amount of savings in the long run.

Installing solar panels on your RV or car is a huge investment of both time and finances, so it's important to do some homework on solar power systems and mistakes to avoid before making your investment.

## 1) IMPROPER WIRING and poor installation
Improper wiring can make your DC isolators

burst into flames; herefore, you should be very mindful when wiring your solar system. If you have no experience or knowledge about wiring electrical systems, it may be better to hire a certified installer to help in installing the panels for you. This professional will implement the appropriate procedures and safety standards when installing the system. They will also advise you on appropriate size for panels, batteries, and wiring to use after evaluating your roof size and your power demands.

You should also put into consideration the basic configuration of the solar panels and whether to use series or parallel connection. Though connecting your solar panels in parallel may increase the current, it does require the use of thicker wires. Connecting thicker wires is not easy because they are hard to bend, especially when running them around corners. The wires are also expensive. Thus, you should be careful when installing them, especially if you're installing them over long distance, as they do help in reducing voltage drops and mitigate the effects of line loss. Poor installation of the components can also affect the performance of the solar power system.

Some of the rules to follow to ensure proper wiring and installation include:

- Ensure all loads from AC are connected to the inverter output, whereas the DC loads should be connected to the solar charge controller output.

- Low-voltage appliances such as refrigerators should be connected to the battery directly.
- Small DC systems connected to the charge controller don't need a fuse apart from the fuse incorporated at the charge controller. Large DC systems require a fuse at the positive battery terminal end.

You should, at all times, follow the appropriate sequence of plugging in the solar charge controller. The process of connecting and disconnecting wires between the panels, solar charge controller, and the battery bank is indicated in the instruction manual. Failure to follow the sequence may lead to the damage of your load due to the high voltage generated by your solar panels.

## 2) Leaving cables exposed

You should not leave wires exposed in an open place. You should also mate the connectors that are attached to the solar panel, as doing so will help you avoid corrosion on the wires that can affect the performance of your entire solar system. Humidity in the area coupled by intense sunlight will increase corrosion of the wires, and it won't take you long before you start to think of replacing them.

### 3) Not getting **multiple quotes**

Before purchasing your solar panel kit, you should look for multiple quotes and compare their pricing. Some companies sell them at four times the price of the total cost of the panels and the installation. Many solar panel companies are in competition with each other, meaning that some may lower the prices of the panels by compromising the quality of the panels.

The cost and durability are some of the factors people have used to compare, so having several choices is a good idea. Get solar estimates from several companies and let them break down the solar cost for you, so you can at least understand what you're buying in consideration of your roof and current power needs. However, relying on too many quotes may become confusing. Some companies may offer sugar-coated information just to make a sale, so do the proper homework before buying. You should also consider buying the solar components from the same supplier because it may be cheaper than buying each component individually.

.   .   .

## 4) BUYING **cheap solar panels**

Buying a cheap solar panel may seem like the right option for you, but remember that you will only benefit for a short time, and it will probably end up costing you more in the future. You may end up spending more money on fixing faulty systems and sometimes you will be forced to replace the entire solar power system.

Cheaper solar panels also have low efficiency. As a result, they may not make enough sizing for your power demands, and you could be forced to install more panels to obtain enough power for your needs. You may also not have enough roof space to install them. You need to invest in quality solar panels; although, this doesn't mean buying the most expensive of all. You can get a professional solar installer to recommend the best solar panels for your needs. For example, if you need a small solar panel for your RV or car, you can buy a quality solar panel at reasonable prices rather than buying a panel from an expensive brand. Depending on your budget, always go for the highly rated panels. The recommended solar panels rating is between 15% and 20%.

## 5) NOT UNDERSTANDING **solar power dynamics**

WHEN INVESTING IN SOLAR PANELS, you have to get

your calculations right. Not doing the right calculations on the panel size may cause the battery to not help you in saving on costs or meeting your power needs. Your goal is to have a solar power that meets the power demands for your appliances at any time even during periods when there is no sun to generate enough power input. This all means that you have to do proper calculations for your off-grid solar system.

## 6) WRONG SYSTEM **size**

One of the common mistakes customers make is installing an oversized system. They will have solar panels that produce more energy than they need, and the extra energy produced ends up going to waste. Buying large solar panels in terms of power output can be very expensive, so there is no reason for you to pay for the power you won't end up using. Using batteries to store excess energy and drawing energy directly from the grid will help you reduce the size of your solar power system. You need to balance the amount of energy used directly from the panels with that which is already stored in the batteries. The batteries make the energy less efficient, so connecting to the grid power will make it easier to utilize the energy generated by the large solar systems and be able to meet your average energy demands.

Another mistake is assuming the solar power rating will be equivalent to how much power output is being generated every day. For example, if the system indicates 6kW, it is only indicating the peak power output of the system. Your system may produce half or more of that based on weather conditions and other factors.

### 7) Battery discharge

Another mistake is discharging the battery beyond 50%. To extend your battery's lifespan, you will have to ensure the batteries do not discharge past 50% of their capacity on a regular basis. Some modern batteries, like lithium-ion batteries, can discharge beyond 90%, putting your batteries at risk. If you maintain the batteries at the recommended discharge level, your battery can last up to 10 years. If your battery discharges more quickly, it will increase your cost of solar power, making you start probably regretting switching to solar energy.

Even if you buy expensive batteries designed to last long, if you don't monitor the discharge on a regular basis, then they will discharge more quickly with time. Batteries have chemical reactions, and if the reactions are left for long in one direction, they become unbalanced and cause problems later. Always buy the right solar battery size for your RV installations and remember that not every solar battery will perform well with your car or RV. Each battery type has its own advantages and disadvantages.

## 8) Not mounting **inverter under shade**

Your inverter should be mounted in a shaded area; otherwise, mounting it in an open area near heat will make it deplete its capacity faster than the normal rate, possibly reducing its lifespan.

## 9) Not practicing **safety standards**

Many people think a 3kW solar system is small, but it can still generate power that can be catastrophic to an individual.

Avoid installing your solar power system in a rush. Solar panels can catch fire and are not properly installed. You should invest in circuit breakers for protection. Having your solar power system run through the circuit breaks is a nice move to avoid any damage because it protects your appliances from overcurrent or overvoltage. In short, solar energy should be treated as live electricity, and care should be taken when handling the solar power components and when installing them.

## 10) Not checking **reviews for solar panels**

Before buying a solar panel kit from any manufacturer, you should evaluate online reviews of the company. The reviews will let you know the credibility of the solar company and whether other customers are satisfied with their products. You will also get to know what other past customers have to say about that product's brand.

## 12) MIXING different solar panels

Mixing solar panels of different watts, voltage, and manufacturers is another common mistake that people make. Though mixing them is not necessarily prohibited, it's not recommended because different solar panels have different power degrade rates. As a result, it may affect the overall efficiency of your solar panels.

Each panel output degrades differently with time, and it is also difficult to get matching panels from different manufacturers. Mixing the solar panel may reduce energy production by 50%, and the amount of drop depends on whether the panels are connected in series or parallel. The solar controller can also be affected by mixing different solar panels with different characteristics. An alternative is to use different controllers for different solar panels, but this will be very expensive for you and can cause clutter in the panels' wiring.

· · ·

## Chapter Summary

Installing solar power systems doesn't necessarily require you to be an expert. The above mistakes can be avoided if you follow the instructions for handling each solar component. Take your time and do your homework in terms of solar panel installation. It's best to spend some weeks figuring out about the solar system suitable for your needs than to invest in a system that will leave you with more regrets. Though getting several quotes can be tiresome sometimes, it will pay off in the long run.

Batteries can last a long time if you don't discharge them beyond the 50% mark regularly. You can install a battery monitor to provide live readings of your battery. Modern lithium batteries are very expensive compared to lead-acid. Though they perform better and are more tolerant to deep discharges, if you don't maintain proper care on them, they will start to discharge faster. All you need is to ensure the battery doesn't discharge beyond the 50% power capacity.

## FINAL WORDS

Just like with solar-powered homes, solar-powered cars extract energy from the sun and convert it to electricity for use with AC and DC appliances. The electricity is generated from the solar panels, then fuels the battery used in running the car.

Solar panels generate electricity through the use of photovoltaic cells (PV) made of semiconductor material. The material absorbs light, leading to the flow of electrons. As the electrons flow, electricity is produced. The generated electricity is used to power batteries known as solar batteries, or into a specialized car motor to run electric cars directly.

Solar energy is a clean, reliable, and renewable source of energy. Solar power is more efficient and helps reduce utility bills. Though the initial cost of installing solar panels is very high, in the long run, it will be one of your best investment decisions. If you love camping or boondocking, solar power will save you a lot. You will be able to camp in any place of your choice when you rely on electricity from the sun.

To benefit from solar power, you will have to build a reliable 12V system. Building a solar power system requires you to list the various solar power components you need to build your solar power system. These major components include: solar panels, a solar controller, inverters, a battery bank, connection cables/wires, fuses, and solar racking equipment. Solar racking is a tool used to mount solar panels securely on the roof of your RV. There are various racking equipment available in the market that you can use, though the decision will depend on the type of the solar panels you have, along with the roof style. After listing the various solar components and functions for each, the next step will be to plan and design your solar system.

Planning involves determining your power requirements. To know your power requirements, you have to list both the DC and AC appliances you need. Each appliance has a label on the total power consumption. You can total the power for all the appliances to figure out your power demands per day, which will help you know the number of solar panels you need to satisfy all your power require-ments. You learned through this book about the process of wire sizing, the different types of wires you can use, how to determine the appropriate wire size for connecting various components, and judging the length of each wire.

The battery capacity and type is also very important. When buying for the battery, make sure to go for solar batteries designed for use in RVs, cars, trucks, and vans. Getting a battery meant for home installation will not work for your RV installation.

You also need a solar charge controller. You can choose between pulse width modulation (PWM) and maximum power point tracking (MPPT). The most common charge controller is MPPT because it's more efficient than the PWM, though it will cost more to buy. Solar charge controllers are designed to use different power levels, so when buying the solar controller, make sure you buy one that can handle the power generated by your solar panels. As a rule, you should have a solar charge controller that has more amps than your solar panels and battery combined. This will be more advantageous when planning to expand your solar system in the future.

After sizing all the required components, you have to figure out how to install each component. In this tutorial, you learned how to install solar panels on the roof, how to install solar controllers, batteries, the battery monitor, and inverters to convert DC to AC. You also learned how to run the cables connecting from point A to point B, use fuses in your wiring, and install circuit breakers to protect your wiring of the various components from any short circuit.

Installation of solar power systems in trucks, cars, and vans follows the same procedure as that of RVs. You learned how to install the solar panels in your boat using different methods. After installing all the components in place, you can then add more DC and AC appliances to your solar RV and have a comfortable and fantastic camping experience.

Proper care and maintenance of your solar system is essential to extending solar panels' lifespans. In this book, you learned how to maintain various solar components and safety standards to practice when operating with them. In the last chapter, we talked about various mistakes people make when handling solar power systems and how you can avoid those mistakes.

Just like with solar-powered homes, solar-powered vehicles can extract energy from the sun and convert it to electricity for use with AC and DC appliances. The electricity is generated from the solar panels, then it fuels the battery used in running the car. The inverter connected allows conversion of DC to AC to make it easy for you to connect your smart devices.

REFERENCES

Basic parts of a DIY camper solar setup. (n.d.).
*Explorist.life DIY Campers.* https://www.ex-
plorist.life/basic-parts-of-a-diy-camper-solar-
setup/

Beasley, K. & Wendler, M. (2020). Best portable
solar panels in 2020. *Camp Addict.* https://campad-
dict.com/portable-solar-panels/

Conger, C. (2008). How can solar panels power a
car? *How Stuff Works.* https://auto.howstuffworks.-
com/fuel-efficiency/vehicles/solar-cars.htm

How much solar power do I need for my RV?
(2019). *Driving Life.* https://drivinglife.net/how-
much-solar-power-for-rv/

How to design solar pv system. (n.d.) *Leonics.*
http://www.leonics.com/support/article2_12j/arti-
cles2_12j_en.php

GSES India Sustainable Energy (Ed.). (2016)

Installation, operation & maintenance of solar PV microgrid systems - A handbook for trainers. *iSolar Alliance.* http://isolaralliance.org/docs/Microgrid-Trainers-Handbook.pdf

Marsh, J. (2019). Solar racking: What you need to know. *EnergySage.* https://news.energysage.com/solar-racking-overview/

petesrv. (2014). 12 volt dual battery setup - Pete's RV quick tips (CC) [Video file]. *YouTube.* https://www.youtube.com/watch?v=9g1xLnFsob0

Pursel, B. (2019). Guide to RV solar panels. *RV With Tito.* https://www.rvwithtito.com/download/RVwithTito-SolarPanelGuide-2019.pdf

saamok. (2020). The 5 best budget solar car battery charger: Why they best. *Green Solar.* https://www.solarpanel.wiki/solar-car-battery-charger/

/1496599699185/OffGridworkshopParticipant-Guide.pdf%20https:/solarquotesnow.com.au/common-mistakes-solar/

ZHCSOLAR. (2019). RV solar controller ultimate guide. *ZHCSOLAR.* https://zhcsolar.com/rv-solar-charge-controller-guide/

Wendler, M. (2020). How to easily design and install RV solar - Part 2: Installation. *Camp Addict.* https://campaddict.com/rv-solar-system-installation/